I0120276

Alexander Harvey

On the Foetus in Utero,

as inoculating the maternal with the peculiarities of the paternal organism, in a

series of essays now first collected

Alexander Harvey

On the Foetus in Utero,
as inoculating the maternal with the peculiarities of the paternal organism, in a series of essays now first collected

ISBN/EAN: 9783337218430

Printed in Europe, USA, Canada, Australia, Japan

Cover: Foto ©berggeist007 / pixelio.de

More available books at **www.hansebooks.com**

ON THE

FŒTUS IN UTERO

AS INOCULATING THE MATERNAL WITH THE PECULIARITIES OF THE PATERNAL ORGANISM.

IN A

SERIES OF ESSAYS NOW FIRST COLLECTED.

BY

ALEXANDER HARVEY, M.A., M.D.,

EMERITUS PROFESSOR OF MATERIA MEDICA IN THE UNIVERSITY OF ABER-
DEEN ; CONSULTING PHYSICIAN TO THE ABERDEEN ROYAL INFIRMARY ;
MEMBER AND LATE PRESIDENT OF THE MEDICO-CHIRURGICAL SOCIETY
OF ABERDEEN, AND OF THE HARVEIAN SOCIETY OF EDINBURGH, ETC.

LONDON :

H. K. LEWIS, 136, GOWER STREET.

1886.

TO

SIR ANDREW CLARK, Bart.,

M.D., LL.D., F.R.C.P., F.R.S.,

SENIOR PHYSICIAN TO THE LONDON HOSPITAL,

IN WHICH HOSPITAL, IN 1835,

THE ESSENTIAL PRINCIPLE OF THIS WORK FOUND FULL EXPRESSION,

THIS VOLUME IS INSCRIBED

BY

THE AUTHOR,

IN RECOGNITION OF ATTAINMENTS, PROFESSIONAL, SCIENTIFIC AND PERSONAL,

WHICH HAVE WON FOR HIM THE HIGH PLACE

LONG HELD BY HIM AS A PHYSICIAN IN THIS METROPOLIS,

AND WHICH HAVE, MORE RECENTLY,

ARNED FOR HIM HONOURABLE DISTINCTION AT THE HANDS OF THE QUEEN.

AND, IN GRATEFUL ACKNOWLEDGMENT, ALSO,

OF MANY MUCH-VALUED PROFESSIONAL AND OTHER KIND SERVICES

RENDERED BY HIM TO THE AUTHOR.

" It is a generally received opinion, I believe, that syphilis in its secondary stage, is not communicable directly to either sex from the other—that the disease is not propagated unless there exist an open chancre ; and this accords with my observation. But it appears to me *probable*, that if a previously healthy woman conceive of an ovum tainted by syphilitic virus derived from its father, her system may become inoculated during the progress of gestation, in consequence of the close vascular connection existing between it and herself ; for it has fallen to my lot *to see more than one case* in which a young woman, united to a man labouring under obstinate secondary symptoms, remained *healthy* for *some months after marriage*, but became the subject of the same disease in its secondary form *soon after impregnation had taken place ;* and *I have considered* that, in such a case, the mother derived the disease, *not directly from the father*, but *from the affected infant* which she carried in her womb."— FRANCIS H. RAMSBOTHAM, M.D.—*Medical Gazette*, May 23, 1835.

PREFACE.

THE Essays—five in number — comprised in this small volume, are, virtually, reprints of papers on the subject of *Fœtal Inoculation* contributed by the Author,—the first three of them, in 1849–50, to the Edinburgh *Monthly Journal of Medical Science*, and the fourth, in 1859, to the *Glasgow Medical Journal*. The fifth and last of the Essays was contributed, in 1866, to the *Indian Annals of Medical Science* by the Author's son—Dr. Robert Harvey, Professor of Midwifery in the Medical College of Calcutta, by whose permission it is appended to this collection.

In now again submitting these Essays to the notice of his professional brethren, the Author would observe that, since the *inoculation theory* was first fully ventilated and brought under discussion, one whole generation of medical men has passed away and been replaced by another; and that as to these, it is not perhaps going too far to say, that to a large number of them the theory in question is but little known, or known only in name, with but little or no knowledge of particulars, and that to not a few it is not known at all. For, thoroughly well grounded as the theory now is, and important as well as manifold as, by those conversant with it, its bearings are known to be, it so happens that somehow it has as yet received only a passing notice —or no notice at all, in our text-books on physiology ; while access to the original papers that treated of it and which lie buried in different medical journals, and even in newspapers, is not readily to be had.

Such being the case, while the subject is, unquestionably, alike interesting and important, the Author would venture to give expression to a hope that this reproduction of the papers referred to, and that too in a convenient because a collected form, will be not unacceptable to many members of the profession. As far as he could he has striven to weed the Essays of redundancies and repetitions. But it will be obvious, on a moment's reflection, that this could not have been effectually accomplised without re-casting and re-writing the entire work—a task which broken health would have forbidden him undertaking.

16, Hanover Terrace,
    Ladbroke Square,
        London, *January* 21, 1886.

CONTENTS.

ESSAY FOURTH.

APPENDIX.

ESSAY FIFTH.

INTRODUCTION.

THE first three of the Essays in this Collection, those which appeared in 1849-50, arose out of an article of great interest which was contributed to the *Aberdeen Journal* newspaper in 1849, by Mr. James Macgillivray, of Huntly, a veterinary surgeon in Aberdeenshire. In that article Mr. Macgillivray made it his business not merely to propound the theory of *Fœtal Inoculation*—the idea of which had independently suggested itself to his own mind—but to state also the grounds on which he advanced it. In his view it is only on such a principle that certain phenomena arising out of cross-breeding can satisfactorily be accounted for—phenomena long familiar to persons engaged in the breeding of different kinds of animals.

Struck with the novelty, and, he may add, with the reasonableness of the view presented in that article, based on the intimate vascular connection subsisting between the Fœtus in utero and its mother, the Author kept a firm hold of it, and following it out as best he could and as opportunity offered, he made inquiry about it in all quarters, and as well among cattle-breeders as among his professional brethren. The result was the publication that same year and the year following of the papers referred to. Further, in the hope of engaging the co-operation in this inquiry of the cattle-breeders (a large body in Aberdeenshire) the Author published, in 1851, a pamphlet on the subject, dedicated to the Highland and Agricultural Society of Scotland, and embodying in a popular form the substance of those papers. This pamphlet bore the title of " On a Remarkable Effect of Cross-Breeding."

These papers and this pamphlet served the purpose aimed at in their publication. The theory thus floated was taken in hand by the profession and by cattle-breeders. The subject was kept alive and largely inquired into. A large body of facts was brought together, and the theory, it may now be confidently affirmed, has come to be accepted as a truth—a *law*—in physiology, and one, too, of the highest importance practically.

The Author had for many years looked upon Mr. Macgillivray as entitled to the paternity of this very beautiful theory. And so far he does so still. The idea of it was entirely original with him; and at the time he advanced it, it had no place in any of our physiological works, text-books, or records.

In none, at least as far as known to the Author, save one — the old London *Medical Gazette* for May 23, 1835. Its issue that day contained the report of a lecture on " Abortion," delivered at the London Hospital by Dr. Ramsbotham. In that lecture Dr. Ramsbotham made the very striking announcement of the theory which stands on the page facing the Preface to this book. It speaks for itself as the fullest yet the most concise, the clearest and most explicit expression of the theory that can well be given. Yet it had long lain in that periodical virtually dead and buried. By the merest chance it was that the Author lighted on it there in 1856, twenty-one years after its publication.*

It does not appear whether Dr. Ramsbotham ever did more than give this expression of the theory. If he did, it is nowhere to be met with, and it is certain that nothing came of it. Pity it is for his own sake that, having himself worked out the idea of it, although as only probably true, he did not keep hold of it and look more deeply into it. Had he done so he could scarcely have failed to see the nail straight ahead of him, and so as to have driven it home—that is, he could scarcely have failed to see the actual truth of the theory; nor this alone, but to see it also as a truth of no mean importance and of wide application in physiology and pathology. This much it is due to the memory of Dr. Ramsbotham (a man highly esteemed in his day) to take note of in connection with this sketch of the evolution of the theory.

In now presenting the papers referred to (the

---

* Dr. Montgomery, of Dublin, refers to it in the second edition of his great work, published in 1856, not in the first, published in 1837. But he does not say when he came to the knowledge of it.

pamphlet on Cross-Breeding alone excepted) in a collected form, under the name of Essays, to his professional brethren, the Author desires it to be understood that he claims no credit in respect of the theory of *Fœtal Inoculation* beyond that of calling attention to it. He did not suggest the theory, and (as will be seen in these pages) he has uniformly disclaimed the paternity whenever ascribed to him by correspondents. As already observed, he only floated it out into the world—the professional world and the agricultural. This credit, however, he does claim, and he deems it credit enough. As to who it is that is entitled to claim the paternity, or rather, in behalf of whom that claim ought in fairness to be set up, whether in behalf of Dr. Ramsbotham or of Mr. Macgillivray—both of them long since dead—or of any one else rather than of either of these, there may be difference of opinion. But, indeed, as regards controversial discussions on the pretensions of different observers to priority of discovery, the Author cordially concurs in a remark made by Dr. Alison—to wit, that "a much greater and more enduring interest attaches to the determination of truth, than to the adjustment of any such personal claims." As to the two observers just named, however, it is clear that from the facts before them they both arrived, independently, at identical conclusions, the only difference between them being one as to time—the one being fourteen years in advance of the other. If there be any other difference it lies in the *kind* of the facts from which they severally deduced their common conclusion, resting it, the one on pathological, the other on physiological facts.

In this brief Introduction the Author, for the most part at least, has confined himself to his own particular relations to the subject of these Essays, and to the circumstances under which those of them that were contributed by himself were written. In the Essays themselves reference is made and full justice done (it is hoped) to those—and these not a few—that have aided in the evolution of this theory, or, rather, this LAW of FŒTAL INOCULATION.

FROM

THE EDINBURGH MONTHLY JOURNAL OF MEDICAL SCIENCE,

For *October*, 1849.

ESSAY FIRST.

ON THE FŒTUS IN UTERO, Etc.

INSTANCES are sufficiently common among the lower animals where the offspring exhibit, more or less distinctly over and beyond the characters of the male by which they were begotten, the peculiarities also of a male by which their mother had at some former period been impregnated,—or, as it has been otherwise expressed, where the peculiarities of a male animal that has once had fruitful intercourse with a female, are more or less distinctly recognized in the offspring of subsequent connections of that female with other males.* A young chestnut mare, seven-eighths Arabian, belonging to the Earl of Morton, was covered in 1815 by a quagga, which is a species of wild ass from Africa, and marked somewhat after the manner of the zebra. The mare was covered but once by the quagga; and, after a pregnancy of eleven months and four days, gave birth to a hybrid which had distinct

---

* Alison, Outlines of Physiology, 3rd. Ed., p. 443.

marks of the quagga, in the shape of its head, black bars on the legs and shoulders, &c. In 1817, 1818, and 1821, the same mare (which had in the meantime passed into the possession of Sir Gore Ouseley), was covered by a very fine black Arabian horse, and produced successively three foals, all of which bore unequivocal marks of the quagga.* Several other examples illustrative of the general fact above stated will presently be given.

Great difficulty has been felt by physiological writers in regard to the proper explanation of this kind of phenomena. They have been ascribed by some to a permanent impression made somehow by the semen of the first male on the genitals, and more particularly on the ova, of the female; and by others to an abiding influence exerted by him on the imagination of the female, and operating on her mind at the time of her connection subsequently with other males, and perhaps during her pregnancy. But they seem to be regarded by most physiologists as inexplicable.

Very recently, in a paper published in the "Aberdeen Journal,"† an intelligent veterinary surgeon, Mr. James M'Gillivray, of Huntly, has offered an explanation, which seems to me to be the true one. His theory is set forth in the following statements quoted from that paper:—"When a pure animal of any breed has been pregnant to an animal of a different breed, such pregnant animal is

---

* Philosophical Transactions, 1821, p. 20; Dunglison's Human Physiology, 3rd. Ed., vol. ii. p. 387.

† March 21 and 28, 1849.

a cross ever after; *the purity of her blood being lost,*
in consequence of her connection with the foreign
animal;" and again: "If a cow, say of the pure
Aberdeenshire breed, is in calf to a bull of the short-
horn breed (known as the Teeswater breed), in proportion as this calf partakes of the nature and physical characters of the bull, just in proportion will the
blood of the cow become *contaminated,* and herself a
cross, for ever incapable of producing a pure calf of
any breed." "It is maintained, therefore (Mr.
M'Gillivray adds), that the great variety of non-
descript animals to be met with are the result of the
crossing system; the prevailing evil of which is
the admission of bulls of various breeds to the
same cow, *whereby the blood is completely vitiated.*"

In explanation of his theory, Mr. M'Gillivray
enters into particulars as to the nature of the connection subsisting between the fœtus in utero and
its mother, with the view of showing (what seems
to him essential to the validity of the theory) that
there is a *direct* vascular communication between
the two; and that, while a portion of the mother's
blood is continually passing by direct transmission
into the body of the fœtus, the latter returns to the
former so much of that blood as is not needed by it,
and that this superfluous blood, after circulating
through the system of the fœtus, passes as directly
into the system of the mother, and, commingling
with the rest of her blood, *destroys its purity, contaminates, vitiates it.*

Mr. M'Gillivray is quite wrong, I apprehend, in
assuming that there is, in any case, a direct vascu-

lar connection between the fœtus and its mother.
Nor does the assumption appear to me at all necessary to establish the theory.   But waiving, for the
present, all discussion of that point, it may here be
observed that Mr. M'Gillivray regards the influence
exerted by the male on the female animal, through
the medium of the fœtus, as *constitutional*; and
perhaps the best general expression of the theory
is, that the fœtus, partaking, as it must, of the
characters or peculiarities of its father, *inoculates*
therewith the blood, and, generally, the system, of
its mother.

The subject now opened up is certainly one of
great interest in general physiology, as well as of
considerable practical importance to breeders.   It
cannot but be interesting to inquire whether the
fact instanced in Lord Morton's mare, is or is not a
general law in animal physiology ; and, if it be,
whether and how far it is modified in its operation,
in different animals, and under different circumstances.   But to the human physiologist, and to
the physician, it is of more immediate interest to
inquire whether or not the fact extends also to his
own species ; and, if it does, to ascertain how far
it applies, and whether it does not admit of illustration by, and serve itself, in its turn, to illustrate
certain known facts in regard to, the communication
and the constitutional effects of the syphilitic and
other morbific poisons, the scrofulous diathesis, &c.
And, in particular, it can hardly fail to suggest some
such curious questions as the following, viz. :—

1st, Whether, in the case of a woman who has been twice married, and borne children to both husbands, the children borne to the second husband ever, or generally, partake of the peculiarities of the first husband.

2nd, Whether, in a family of several children, the younger children, rather than the elder, are disposed, *cæteris paribus*, to exhibit the characters of the father.

3rd, Whether a woman who has borne several children by the same husband, may not ultimately acquire some of the physical characters, or at least imbibe and manifest some of the morbid tendencies, of the latter.

In treating further of this singular subject, I shall *first* state the facts at present known to me regarding it; and, *secondly*, consider the theories offered in explanation of it.

I. In regard to the facts of the subject, these will be most conveniently noticed, *first* in relation to the lower animals; and, *secondly*, in relation to the human species.

(1.) As regards the brute creation :—Besides the instance already quoted of the mare belonging to Lord Morton, there is another similar case recorded. A mare belonging to Sir Gore Ouseley was covered by a zebra, and gave birth to a striped hybrid. The year following, the same mare was covered by a thoroughbred horse, and the next succeeding year by another horse. Both the *foals* thus produced were striped, *i.e.*, partook of the character of the zebra.* And it is stated by Haller, and also by Becker, that when a mare has had a *mule* by an ass, and afterwards a *foal* by a horse, the foal exhibits traces of the ass.†

In the foregoing cases, the mares were covered, in the first instance, by animals of a different species from themselves. But cases are recorded of mares covered in every instance by horses, but by different horses, on different occasions—where the offspring partook of the characters of the horse, by which impregnation was first effected. Of this Mr. M'Gillivray gives two examples. Thus, in several foals, in the Royal stud at Hampton Court, got by the horse *Actæon*, there were unequivocal marks of the horse *Colonel*: their dams, however, had been covered by Colonel the previous year. Again, a colt, the property of the Earl of Suffield, got by *Laurel*, so resembled another horse, *Camel*, " that it was whispered, nay, even asserted, at Newmarket, that he must have been got by Camel." It was ascertained, however, that the mother of the colt was covered, the previous year, by Camel.

It has often been observed, also, that a well-bred bitch, if she have been impregnated by a mongrel-dog, will not, although lined subsequently by a pure dog, bear thorough-bred puppies in the next two or three litters.‡

---

* M'Gillivray, " Aberdeen Journal," March 28, 1849. Paintings of these animals and their skins are said to be preserved in the Museum of the Royal College of Surgeons of England.

† Haller, Element. Physiol., viii. p. 104 ; Becker, Physic. Subterran. Lips., 1703. Quoted from Dunglison's Physiology, vol. ii. p. 387.

‡ Kirkes' Handbook of Physiology, p. 613.

The like occurrence has been noticed in respect of the sow. A sow of the black and white breed (known as Mr. Western's breed) became pregnant by a boar of the wild breed, of a deep chestnut colour. The pigs produced were duly mixed, the colour of the boar being in some very predominant. The sow being afterwards put to a boar of the same breed with her own, some of the produce were observed to be stained or marked with the chestnut colour that prevailed in the former litter. And, on a subsequent impregnation, the boar being still of the same breed as the sow, some of the litter were also slightly marked with the chestnut colour. What adds to the value of the fact now stated is, that in the course of many years' observation, the breed in question was never known to produce progeny having the smallest tinge of the chestnut colour.*

Breeders of cattle are familiar with analogous facts as occurring in the cow. A pure Aberdeenshire heifer was served with a pure Teeswater bull to whom she had a *first-cross* calf. The following season, the same cow was served with a pure Aberdeenshire bull ; the produce was a *cross* calf, which at two years old had very long horns, the parents both

---

* Philosophical Transactions for 1821, p. 23. "Apart from a state of domestication," says Mr. M'Gillivray, "I do not believe that there is one solitary instance in which an animal has produced offspring of various colours. Animals, left entirely to the operation of natural causes, never exhibit this sporting of colours ; they are to be distinguished by various and often beautiful shades of colour ; but then each species is true to its own family type, even to a few hairs or small parts of a feather."

hummel.*   A pure Aberdeenshire cow was served, in 1845, with a cross bull—*i.e.*, an animal produced between a first-cross cow and a pure Teeswater bull. To this bull she had a cross calf.   Next season she was served with a pure Aberdeenshire bull—the calf was quite a *cross* in shape and colour.†

Mr. M'Gillivray, after narrating the whole of the foregoing examples, says :—" Many more instances might be cited, did time permit.   *Among cattle and horses they are of every-day occurrence.*"

(2.) As regards the human species.   The facts bearing on this division of the subject are exceedingly few, and not to be relied on ; and the observations which follow are intended rather to suggest and direct, than to satisfy, inquiry.

Dr. Allen Thomson, in his article on Generation, in the " Cyclopædia of Anatomy and Physiology," remarks :—" It is affirmed that the human female, when twice married, bears occasionally to the second husband, children resembling the first, both in bodily structure and mental powers."   And Dr. George Ogilvie, of this city, informs me of a case, which fell under his own observation, where a woman was twice married, and had children by both husbands, and where the children by both marriages were scrofulous, although only the first husband had marks of that diathesis ; the woman herself, and her second husband being to all appearance quite healthy.

Dr. Ogilvie's case is clearly beset by so many sources of fallacy, that we cannot venture at pre-

---

* M'Gillivray, loc. cit.   † M'Gillivray, loc. cit.

sent to regard it as a case in point. Dr. Thomson
does not bring forward any instances, nor offer any
proof, in support of his statement; and, indeed, he
gives it, without saying whether he thinks it well
or ill-founded. Any such statement, it is plain,
based on observation of the children of European
parents—*i.e.*, where the female and both her husbands and their children are all white—must be
comparatively difficult of verification; but it is
equally plain that means exist for subjecting it to a
pretty decisive test. There are equally distinct
breeds of the human family as of any of the lower
animals; and all that seems requisite in regard to
determining the question under consideration is,
to observe accurately whether the children of
European parents, where the woman has, in the
first instance, had offspring by a negro, exhibit
traces of the latter in the colour of the skin, the
form of the features, &c.; or, *vice versâ*, whether
the children of negro parents, where the woman
had, first of all, been impregnated by a European,
exhibit the peculiarities of the latter. Of the
former case, a medical friend informs me that he
recollects hearing of an instance of the kind as
occurring in this neighbourhood, but cannot vouch
for the truth of it. Of the latter case, if the
general fact applies to the human species, instances
must abound in our West India colonies, in the
United States of America, and in other parts of
the world. My colleague Dr. Dyce tells me, that
he has certainly known one instance (if not more)
where a creole woman bore fair children to a white

man; and that the same woman had afterwards to a creole man other children, who bore much resemblance to the white man, both in features and in complexion. But two very intelligent friends—the one a West India proprietor, the other a medical man—both long resident in Jamaica, tell me that they never noticed, nor ever heard of an instance of the kind, although connections of that sort are common there, and children born under such circumstances very numerous. It is singular, indeed, if instances of the fact in question do occur, and still more, if they are of frequent occurrence, that they should not be notorious. It is conceivable, however, and by no means improbable, that cases do exist, but that they have been overlooked from the traces of the European being so minute as to escape ordinary observation, and that the fact has remained unknown from special attention never having been directed to it.

If the male does exert any such influence as is here in question on the constitution and the reproductive powers of the female, it is conceivable that, by each successive impregnation effected by him, that influence may be increased; and, if so, the younger children begotten by him, rather than the elder, might be expected, *cæteris paribus*, to bear their father's image. And this more special fact, if ascertained, would establish also the more general one. I am not aware, however, of any specific facts bearing upon it, nor of any popular notions regarding it. But my colleague Dr. Laing is cognizant of the case of an English gentleman who

had a large family by a negro woman, in the West
Indies, and where the children successively exhibited more and more the European features and
complexion.

But, however this may be, there is a popular
belief that, in the course of years, a woman comes
to resemble her husband, and that not merely in
respect of temper, disposition, or habits of thought,
but in bodily appearance. But, in as far as the
notion may hold good, it may be true only of the
features, and of these only as they indicate or be-
speak the inward feelings of the mind, which, from
long and familiar intercourse, may, to a certain
extent, become common to man and wife. In as
far as the notion is true in any other respect, and
the parties have had several children, it may suggest the question, whether the assimilation is not
referable to an influence exerted by the husband,
through the medium of the fœtus in utero, on the
constitution of the wife ? The question is probably
an idle one, and the notion only a popular error.
In as far, however, as there is anything in it, the
explanation suggested gives a peculiar, and it may
be added, a physiological significancy to the lan-
guage of Scripture relative to man and wife, at
least when their intercourse has been fruitful—
" They twain shall be *one flesh.*"

It is of more immediate interest, however, and of
greater practical moment, to ascertain whether,
through the medium of the fœtus, the husband
may impart to his wife either the syphilitic virus,
or the scrofulous diathesis, or any other constitu-

tional morbid tendency (*e.g.* insanity) which he
may possess. Facts are wanting on this subject;
but it is not undeserving of patient inquiry. Dr.
Ogilvie's case, formerly referred to, if it could be
relied on, would be an instance of it. Before the
mother could have imparted the scrofulous taint to
her offspring by the second husband, she must herself have imbibed it from her first husband through
the medium of his offspring while in utero. And,
although still seemingly free of the taint, it may
have required only the appropriate external conditions to call it into full activity in her own person. And, with regard to the syphilitic poison,
there is no difficulty in understanding, and it is
quite within the bounds of probability, that the
fœtus, if contaminated with it by its father, may
convey it to the mother. Messrs. Maunsell and
Evanson, after mentioning that they have notes of
the case of a syphilitic child, whose mother had
been infected by a former husband (they do not say
in what way)—and to all appearance cured five
years before its birth—the father of the child (her
second husband) being in good health, state that
their experience would enable them to adduce many
curious facts bearing on the communication of the
syphilitic poison.* Perhaps their experience might
furnish an affirmative solution of the question at
issue. It has been affirmed, indeed, that a man
who has once had syphilis, but been seemingly
cured of it for many years, may yet so retain the

---

* On the Management and Diseases of Children, 5th edit. pp.
452–3.

taint of it as to contaminate his offspring, without, at the same time, tainting his wife. Very possibly. But this does not prove that he may not contaminate his wife also ; and the observation itself is in that respect fallacious, inasmuch as in any given case of the kind, the wife may really have imbibed the virus, although in a latent form, and might subsequently give proof of the reality of the fact by tainting the offspring begotten by another and a perfectly healthy husband. Adopting this view, it may be found of importance, in contemplating marriage with a widow, to inquire into the constitutional peculiarities of her deceased husband.

II. Of the general fact now under consideration, and clearly established in respect of the lower animals, only two explanations, that are at all rational, have been offered. The first is that suggested by the great Haller, who ascribes it to a permanent impression made by the semen of the male on the genitals, and more particularly on the ova, of the female ; the second, that suggested by Mr. M'Gillivray, who ascribes it to an influence exerted by the fœtus in utero on the constitution of the mother. The notion entertained by Sir Everard Home and others, that it is an affair of the imagination, seems too absurd to require serious consideration.

Haller's knowledge of 'the subject appears to have been very limited, and his explanation of it to have been offered incidentally. He was aware that when a mare has had a mule by an ass, and after-

wards a foal by a horse, the foal exhibits traces of
the ass; and he remarks, "that the female organs
of the mare seem to be corrupted by the unequal
copulation with the ass,"* *i.e.*, that the semen of the
latter exerts an influence on the genitals, and, of
course, on the ova of the mare, which appears subsequently on the impregnation of these ova by
males of her own species.†

It may be stated, in support of Haller's theory,
that, in the case of birds, a single intercourse is
known to impregnate many eggs which are laid
successively after it; but, on the other hand, the
influence of such intercourse extends only to the
eggs of one season, or rather of one brood,—the
several eggs being laid in tolerably quick succes-
sion, and all of them probably in a state of maturity,
and actually impregnated at the time of that intercourse. This fact, therefore, goes but a short way
to favour Haller's theory, and may, indeed, be said
to tell as much against as for it. And if it shall
be clearly ascertained (as seems presently to be the
belief of physiologists), that any single ovum remains but a short time in the ovary, Haller's theory
must be given up. But even if it could be shown
that an ovum may remain in the ovary for a series

---

* Dunglison's Physiology, vol. ii. p. 387.

† Dr. Kirkes also appears to regard the cause as a local one.
Referring to Lord Morton's case, he observes,—"The single impregnation, by the seminal fluid of the quagga, had impressed its
character not only on the ovum then impregnated, but on the three
following ova impregnated by horses."—Handbook of Physiology,
p. 614. Such, too, is Mr. Mayo s view. Physiology, 2nd Ed.,
p. 490.

of years, the fact would be of little value, unless it could also be shown that the semen can exert some definite kind of influence on an ovum, which it does not at the time actually impregnate. There seems little probability, however, of this being done; and there is one fact known in regard to the ova which makes it difficult to conceive it possible —the fact, namely, that unripe or immature ova lie deeply imbedded in the stroma of the ovary.

Mr. M'Gillivray's theory seems to me to meet the whole facts of the case, and to derive support from a great variety of facts in regard to the reception and constitutional effects of morbific poisons and morbid diatheses.

Mr. M'Gillivray, indeed, supposes, as was formerly noticed, that there is a *direct* vascular connection between the fœtus in utero and its mother; and he seems to consider the validity of his theory to hinge on this assumption. The assumption, however, is untenable, nor is it at all necessary for the establishment of the theory. The researches of Dr. John Reid and of Mr. Goodsir, on the structure of the placenta, have demonstrated that the connection is *indirect* only—the fœtus and the mother imbibing materials from each other, very much in the same way that the lacteal vessels take up the nutritive portions of the food in its transit along the small intestines; or that the roots and leaves of vegetables take up nourishment from the soil and the atmosphere,—the materials imbibed, in each case, passing through a pervious, but not a perforated, tube or membrane, and being taken up by

a real act of *absorption,* during which act they are
more or less altered in their character, or *assimi-
lated.* But, independently of the considerations
now stated, it appears from the observations of
Prevost and Dumas, and of others, that the corpuscles of the fœtal blood are differently shaped
from, and, in the later stages, larger than those of
the mother*—a fact which shows, at least, that no
entire corpuscles of blood are transmitted from the
one to the other, and, indeed, taken in connection
with the facts ascertained as to the structure of the
placenta, proves that it is by *transudation* only that
the contents of the uterine and fœtal vessels mutu-
ally pass into each other.

In so doing, the materials in question are more
or less altered in their character, or undergo what
physiologists term a process of *assimilation.* In
the case of the lacteal vessels, the chyle which
they contain can never be detected *as such* in the
alimentary mass; nor is the sap of vegetables
precisely the same fluid that exists in the soil and
in the air. In like manner, the blood in the umbilical vessels doubtless differs from that existing
in the uterine sinuses. At the same time, the
assimilating process does not go the length probably
in any case of wholly changing the character of the
fluids concerned in it; and there is reason to
believe, that, in different cases, it proceeds to a
very different extent—in some the change effected

---

* Alison's Outlines of Physiology, 3rd. edit. p. 426.—Kirkes'
Handbook, pp. 66-7.

being to a less extent than in others. And possibly, in the case of the fœtus and its mother, the amount of the assimilation is not considerable. No interchange of corpuscles takes place, but, in respect of the other constituents of the blood, it is difficult to conceive why they should not be transmitted nearly unchanged. Professor Simpson (the late Sir James), of Edinburgh, has recently shown that the smallpox virus may pass unaltered from the mother to the child in her womb, and produce in it the actual disease, even although, by reason of previous vaccination, the mother may herself remain unaffected by it.* And a similar fact has long been known in regard to the transmission of the syphilitic virus from the mother to the fœtus in utero.

We can, therefore, have no difficulty in understanding, in respect of the fœtus itself, that, although its connection with the mother is indirect only and merely to the extent of allowing the passage of the *liquor sanguinis*, and although this may even be so far altered in the passage, the constitutional peculiarities, derived to it from its father, and inherent in its blood, may, with the blood, be imbibed by its mother. And when we reflect on the length of time during which the connection between them is kept up, the amount and the activity of interstitial change continually going on in the system of the fœtus, the large quantity of fœtal blood that must eventually be taken into the vessels of the mother, and the probability that the

---

* Edinburgh Monthly Journal of Medical Science for April, 1849.

peculiar matter imparted by the male parent to the ovum at the moment of impregnation (be its nature what it may, and its quantity never so infinitesimal), assimilates, like a ferment, much of the fœtal blood to itself, it does not seem too hard to be believed that the blood and constitution generally of the mother may, thereby, become so imbued with the peculiarities of that parent, as to impart them to any offspring she may subsequently have by other males.

Aberdeen, *April*, 30, 1849.

POSTSCRIPT.

In the foregoing essay a question occurs as to whether, in the case of a woman who has been twice married, and borne children by both husbands, the children of the second marriage ever resemble the mother's first husband (pp. 7 and 10).

The following additional cases, illustrative of this question, have recently been communicated to me : the first by my friend the Rev. Charles M'Combie, of Tillyfour, minister of Lumphanan, in Aberdeenshire; the second by Professor (the late Sir James) Simpson, of Edinburgh; and the third by Professor Pirrie, of Aberdeen :—

1. Mrs. ——, a neighbour of Mr. M'Combie, was twice married, and had issue by both husbands. The children of the first marriage were five in number ; of the second, three. One of these three, a daughter, bears an unmistakable resemblance to her

mother's first husband. What makes the likeness the more discernible is, that there was the most marked difference, in their
features and general appearance, between the two husbands.

2. A young woman, residing in Edinburgh, and born of white
(Scottish) parents, but whose mother sometime previous to her
marriage had a natural (mulatto) child by a negro man-servant,
in Edinburgh, exhibits distinct traces of the negro. Dr. Simpson,
whose patient the young woman at one time was, has had no
recent opportunities of satisfying himself as to the precise extent
to which the negro character prevails in her features ; but he recollects being struck with the resemblance, and noticed particularly
that the hair had the qualities char·cteristic of the negro.

3. Mrs. H——, apparently perfectly free from scrofula, married
a man who died of phthisis. She had one child by him, which
also died of phthisis. She next married a person who was to all
appearance equally healthy as herself, and had two children by
him, one of which died of phthisis, the other of tubercular
mesenteric disease—having, at the same time, scrofulous ulceration
of the under extremity.

APPENDIX.

—o—

On a Singular Result, as regards the Female, of Fruitful Intercourse between the Aboriginal Female of certain Parts of the World and the European Male ; as Observed and Vouched for by the Count de Strzelecki.*

As bearing more or less directly on the subject of this Essay, the Observations following, as to the Result of the intercourse above indicated, seem to me not undeserving of consideration. They clearly point to a *constitutional* influence of some kind as exerted by the Male, through the medium of the Fœtus, on the system of the Female.

The Count de Strzelecki's own statement as to the Result in question is this,—" *Whenever such intercourse takes place, the native female is found to*

---

* The subject matter of this Appendix is taken from that in the author's pamphlet, " On a Remarkable Effect of Cross-Breeding," published in 1851 ; and this, the rather, because of its being more fully treated of there than in the Essay in which it first appeared in 1849.

lose the power of conception on a renewal of inter-
course with the male of her own race, retaining only
that of procreating with the white men." *

Extraordinary as this statement is, it is not
lightly made; nor have the opportunities enjoyed by
the Count de Strzelecki of making observations regarding it been small. He has "lived much (to use
his own words) amongst different races of aborigines,—the natives of Canada, of the United
States, of California, Mexico, the South American
Republics, the Marquesas, Sandwich, and Society
Islands, and those of New Zealand and Australia." †
And, referring to the statement already quoted, the
Count observes—"*Hundreds of instances* of this
extraordinary fact are on record in the writer's
memoranda, *all recurring invariably under the same
circumstances*, amongst the Hurons, Seminoles, Red
Indians, Yakics (Sinaloa), Mendosa Indians,
Araucos, South Sea Islanders, and natives of New
Zealand, New South Wales, and Van Dieman's
Land; and all tending to prove that the sterility of
the female, which is relative only to one and not to
another male, is not accidental, but follows laws as
cogent, though as mysterious, as the rest of those
connected with generation." ‡

In the work in which the foregoing allegations
are made, Strzelecki does not state whether he
has ever met with *exceptional* cases,—that is to say,

---

* "Physical Description of New South Wales and Van Dieman's
Land," p. 347.
† Ibid. p. 345.　　‡ Ibid, p. 347.

cases where, after connection of the kind in question, fruitful intercourse has taken place between a native man and woman. But in a communication on the subject with which he has favoured me, the Count assures me, that he has never met with such a case. " It has not (he writes me) come under my cognizance to see or hear of a native female, who, having a child with a European, had afterwards any offspring with a male of her own race." And I am informed by Professor Goodsir, and by Dr. Carmichael, of Edinburgh, and by Dr. Maunsell, of Dublin, that they have learned from independent sources, that, as regards the aborigines of Australia, Strzelecki's statement is unquestionable, and must be regarded as the expression of a law of nature.*

Assuming that the fact is truly a law of nature, and that it holds as absolutely and extensively as Strzelecki's experience would lead us to infer, " it is," as Professor Goodsir observes, " a very remarkable one, and indicates a series of influences of high import in the natural history of the human race." What that import is, it may not be easy to comprehend, nor perhaps is it a short line that will fathom it. But it seems to indicate, how little account

* "The intercourse of Kamehamha's men [people] and that of the whale ships [manned by Europeans], which now began to anchor in their waters, was sadly disastrous to the native constitution and morals, poisoning the fountains of health, and inducing premature decay and *barrenness.*" This observation, though by no means very definite, seems to point to the fact alluded to in the text. See an article entitled " On the Sandwich or Hawaiian Islands," in the " New York Biblical Repository and Critical Review," for July, 1849.

soever might be taken of it by the author of " The
Vestiges of the Natural History of Creation," that
there is in nature a principle of *degradation*, as well
as a principle of *development*. The fact itself, it
may be observed, is brought forward by Strzelecki
as affording an explanation, and as being the chief
cause of the gradual diminution and ultimate extinction of the native tribes in most parts of the
New World, which follow the introduction of the
European races. " Wherever the white man has
set his foot-mark, there the print of the native foot
is obliterated ; and as the tender plant withers
beneath his tread, so withers the aboriginal in-
habitant of the soil." * And " human interference," says Strzelecki, " to avert this melancholy
consequence has been hitherto of no avail ;—a
charter for colonization granted to one race becomes
virtually the decree for the extinction of the
other." †

---

* Brooke's " Narrative of Events in Borneo and Celebes, ' vol. i.
p. 12.

† Strzelecki records the following remarkable circumstance
which came within his own personal knowledge :—A party of
aborigines in Van Dieman's Land, to the number of 210, were
deported by government in 1835 to Flinders' Island, on account of
aggressions made by them on the colonists in their neighbourhood,
by whom, however, they had been contaminated. They had only
fourteen children born among them during the next seven years.
It is true that, in the course of that time their own numbers had
dwindled away to fifty-four. Still, the small number of births is
singular, and contrasts strikingly with the fact, that " each
family, in the interior of New South Wales, uncontaminated by
contact with the whites, *swarms with children*."—(Op. cit. pp.
350–5.)

Very various causes, doubtless, concur to bring about this result. The one assigned by Strzelecki as the chief is obviously quite adequate, if a real one, to its production. And should his belief as to the reality of this cause be confirmed, and should it further appear, that the principle involved in it applies only to aboriginal females contaminated by European males, and not to European females contaminated by aboriginal males, that is to say, should the former class of females only, and not the latter, be rendered sterile to males of their own race by such foreign intercourse, the discovery can scarcely fail, not merely to exhibit the predominancy of the white over the dark races of men, in a particular not previously suspected, but to indicate that the designs of Providence, in regard to the human family in this stage of existence, embrace the ultimate extinction of the primitive varieties of the dark races, or at least of certain of these. Their physical peculiarities and their social degradation—a mystery, if not a standing memorial of a curse visited on their progenitors, in the times of miraculous interposition; the purpose of their existence in respect of this earth—a mystery also, yet somehow subservient, seemingly, to that of their more favoured *brethren;*—their end, after that purpose is served—extirpation? But these are questions which, besides that they are foreign to the object of this paper, are, perhaps, too deep for human penetration.

It may not, however, be out of place to remark, that it may form part of the plan of Providence

that certain races of men should hold given portions of the earth's surface till certain other races,
and in particular our own Anglo-Saxon race, are
ready to step in and occupy them,—those primitive
races then disappearing; and that the law in question may be directly subservient to the extermina-
tion of these. The rapid increase of the Anglo-
Saxon race during the last two centuries, its wide
diffusion over the globe, and its superiority over
every race with which it has come in contact, are
in harmony with that supposition. It has recently
been stated, with regard to this race, that while in
1620 it numbered only about six millions, and was
almost exclusively confined to the United Kingdom,
it now numbers sixty millions of human beings,
planted upon all the islands and continents of the
earth, and fast absorbing or displacing all the
sluggish or barbarous tribes on the continents of
America, Africa, Asia, and the islands of the sea;
and that, increasing everywhere by an intense ratio
of progression, it is estimated that, if no physical
revolution supervene to check its propagation, it
will number eight hundred millions of human
beings in less than 150 years from the present
time.*

It seems tolerably certain that Strzelecki's law
does not extend to the Negro race—the fertility of
the Negro female with the male of her own race
not being apparently impaired by previous fruitful

---

* American Paper. Quoted from "Chambers's Edinburgh
Journal," for July, 1850.

intercourse with the European male,—a kind of intercourse which is notoriously common in all the West India Islands, the Brazils, and the slave-holding States of North America. But it is yet, I apprehend, undetermined, and it must surely be interesting to ascertain whether the law applies to the Mongolian and pure Malay races, inhabiting China, Japan, Borneo, and the islands of the Eastern Archipelago. And, if it should, one cannot help considering, in connection with it, the footing which Britain has of late acquired in those regions, —the probable rapid increase of our countrymen there and in Australia, Van Dieman's Land, and New Zealand,—and the Anglo-Saxon *nation* rising up as with mushroom growth in California, to over-spread the western coast of America,—and asking whether, through the instrumentality of that law, the native inhabitants of those parts of the East may not yet be rooted out?

But our knowledge of this law must be far wider and more precise than it yet is, to enable us confidently to speculate in regard to it. We do not even know for certain that it holds absolutely over the races specified by Strzelecki, or that the Negro race is wholly exempt from it. It is conceivable, in regard to the former, as Strzelecki himself frankly allows to me, that there may be exceptions to it, though he has met with none; and, in regard to the latter, that it may be in some degree subject to it. It is well known, indeed, that in the West India islands, under the system of slavery at least,

the black population tends rather to diminish than
increase, and that it can only be adequately maintained by continual importations from Africa.
Whether that diminution is in any degree owing to
the operation of Strzelecki's law, it is perhaps at
present impossible to say.

I would only remark further, that in arguing
here from Strzelecki's inference, it is only pro-
visionally, and on the supposition of its being a
fact; and if a fact, then, of course, a *law* of nature;
and that in this, as in every other part of this
Essay, my object is to suggest and direct inquiry,
not to dogmatize. A correspondent, indeed, laughs
at me for having elsewhere given heed to that inference. On the other hand, a writer in the
"Edinburgh Review" (No. CLXXXIV., p. 456,
foot note), referring to one of my papers in the
Edinburgh Monthly Journal of Medical Science,
suggests that Strzelecki should *"excuse"* me for
treating it as nothing more than an hypothesis for
the present! Considering, however, on the one
hand, the astounding nature of the inference, and,
on the other, the high character of Strzelecki as a
philosophic observer, as well as the extent of his
observations (some *hundreds* of instances of the
phenomenon without a single exception), I humbly
submit that, until the inference shall be either fully
established or shown to be fallacious, the proper
course in regard to it is, "to keep the mean between
the two extremes of too much stiffness in refusing,
and of too much easiness in admitting" it,—or (in

the words of the wise and good Bishop Butler) to
keep "in the middle state of mind," "between a
full satisfaction of the truth of it, and a satisfaction
of the contrary"—this middle state of mind consisting "in a serious apprehension that it may be
true, joined with doubt whether it be so."—
(*Analogy*, Part II., Conclusion.)

ESSAY SECOND.

ON THE FŒTUS IN UTERO AS INOCULATING THE MATERNAL
WITH THE PECULIARITIES OF THE PATERNAL
ORGANISM;

AND

ON THE INFLUENCE THEREBY EXERTED BY THE MALE ON
THE CONSTITUTION AND THE REPRODUCTIVE
POWERS OF THE FEMALE—*(continued)*.

D

FROM

THE EDINBURGH MONTHLY JOURNAL OF MEDICAL SCIENCE

For *October*, 1850.

ESSAY SECOND.

ON THE FŒTUS IN UTERO, Etc.

In a paper published last year in the *Edinburgh Monthly Journal of Medical Science*, I brought together a considerable number of instances which serve clearly to show that the peculiarities of a male animal that has once had fruitful intercourse with a female can sometimes be more or less distinctly recognized in the offspring of subsequent connections of that female with other males. I therein stated that of this singular phenomenon two explanations used formerly to be offered,—one class of physiologists referring it to a permanent impression made somehow by the semen of the first male on the ova of the female,—another class ascribing it to an abiding influence exerted by that male on the imagination of the female, and recurring to, and operating on, her mind at the time of her connection subsequently with other males. At the same time I expressed an opinion that neither of these explanations could be regarded as satisfactory; and I

endeavoured to show that an explanation offered by
Mr. M'Gillivray, of Huntly, is the true one,—namely,
that while, as all allow, a portion of the mother's
blood is continually passing by absorption (and
assimilation) into the body of the fœtus, in order to
its nutrition and development, a portion of the blood
of the fœtus is as constantly passing, in like manner,
into the body of the mother;—that as this com-
mingles there with the general mass of the mother's
own blood, it inoculates her system with the constitutional qualities of the fœtus;—and that, as
these qualities are in part derived to the fœtus from
its male progenitor, the peculiarities of the latter
are thereby so engrafted on the system of the female
as to be communicable by her to any offspring she
may subsequently have by other males.

Since the publication of that paper I have learnt
nothing to shake, but much to support, the notion
of *inoculation* therein suggested; and have ascertained that many among the agricultural body in
this district are familiar, to a degree that is annoy-
ing to them, with the facts then adduced in illustration of it,—finding that, after breeding *crosses*, their
cows, though served with bulls of their own breed,
yield crosses still, or rather mongrels; nay, that
they were already impressed with the idea of *contamination of blood* as the cause of the phenomenon,
—possibly, however, on the principle (adverted to
by D'Alembert) that the doctrine so intuitively com-
mended itself to their minds as soon as stated, that
they fancied they were told nothing but what they

knew before,—so just is the observation, that "truth *proposed* is much more easily *perceived* than *without* such proposal it is *discovered.*"*

But with regard to any one of the cases formerly given, as instances of inoculation by the fœtus, I would not now assert that it is an *unequivocal* instance; nor would I now affirm that the notion, entertained by Sir Everard Home and others, of all of them being an affair of the imagination, is so absurd as it once seemed to me. The careful consideration, recently, of cases seemingly identical in kind with these, save and except that there never was coitus between the mother of the animals and the male animal which they resembled, has convinced me that it is not so clear a matter that they may not be referable to the agency of mental causes. A setter bitch, for example, takes a fancy to a cur dog, and, without ever having access to him, produces once and again, to dogs of her own breed, whelps bearing a decided resemblance to the cur. I am still inclined to think, however, that there are good grounds for believing that the greater number at least of the cases are truly referable to the principle of inoculation; and it is very clear that such facts as those exemplified in the progeny of the setter are noways adverse to that theory.

It seems to me probable, indeed, that future inquiry will demonstrate that there are in nature two sets of cases connected with the peculiarities of offspring, analogous, nay, identical, in their external characters, yet essentially different in their origin

* Archbishop Secker, Works (Ed. 1811), Vol. iii. p. 237.

and cause,—the one referable to mental states in one or other of the parents, and oftenest in the female parent,—the other to a change effected in the constitutional powers of the female by a physical (organic) agency originally extrinsic to her, but inherent in a former fœtus by derivation from its male progenitor, and conveyed to her, in the way of inoculation, by that fœtus while in utero.

To establish this, and to determine also the relative proportion of the two sets of cases, together with the comparative efficacy of their respective causes, and the conditions that are essential to the agency of each cause, would be an achievement in physiology of great general interest, and, to breeders, of much practical value. It were idle, perhaps, to engage in so wide and difficult an inquiry with any hope of complete success. But as to some important points, and in particular as to the main question here at issue, experiments in breeding on the lower animals, so conducted as to be free from essential sources of fallacy, might, I am informed, easily be made; and as they could be carried on on the large scale, decisive results might ere long be obtained. Some suggestions as to the kind of experiments which might be instituted, and the mode of con-ducting them, will be offered in the sequel. As to their probable issue, it were, perhaps, wise at present not to hazard an opinion. But, " if reason can sometimes go farther than imagination can venture to follow," the latter can sometimes go in advance of, and pioneer the way to, the former. And possibly in this instance, if, leaving *reason* to follow at her leisure in the broad day and clear sunshine, we hold on our way, warily, in the dim twilight, in company with *imagination*, we may, under her guidance, securely reach a post in advance, and have the satisfaction afterwards of finding reason treading in our steps.

Before entering on the consideration of mental states in either parent, as influencing the nutrition and development of the fœtus in such a manner as to cause the latter to resemble a male animal not its progenitor, I purpose, first, in this second Essay, to submit some additional observations on the theory of inoculation, with the view of placing this doctrine on a firmer footing.

(1.) Without at present raising any question as to the nature of the phenomenon, I may remark, generally, that, since the publication of my former paper, further inquiry has satisfied me of the truth of an observation there quoted from Mr. M'Gillivray, —namely, that cases of the kind given as instances of inoculation of the maternal system by a former fœtus are, as regards *cattle* and *dogs*, " of every day occurrence ; " and judging from the two remarkable cases presently to be given of the like phenomenon in the *sheep*, they are probably equally common in other classes of animals. I have not, indeed, seen any such myself, but opportunities for doing so have not come in my way ; and the information I have received is only of a general kind, sufficient to excite to further inquiry, but wanting in specific details. There is one fact, however, kindly communicated to me by Sir John S. Forbes, of Pitsligo, which, as perhaps bearing on this subject, it seems proper to mention. It is the fact, observable within these few years at all our great cattle shows, that among the high-bred heifers and cows of the Angus polled breed, *fat is laid on in lumps about the tail*, by feeding of the most moderate description—a feature heretofore considered to be characteristic of the Teeswater breed, and which, as occurring in the breed in question, though usually attributed to improved management and high breeding, may, perhaps, have its true solution in Mr. M'Gillivray's theory,—on the supposition that the mother, however pure her own original breed and that of the father, had previously borne offspring to a Teeswater bull.

The following are the cases illustrative of the effect on the procreative powers of the *ewe* of previous sexual intercourse with a ram of a different breed ; the one communicated to me by my friend Dr. William Wells, of the island of Grenada ; the other, by Mr. William M'Combie, Tillyfour, in Aberdeenshire.

1. A small flock of ewes, belonging to Dr. Wells, were tupped a few years ago by a ram procured for that purpose from the manager of a neighbouring estate. The ewes were all of them *white* and *woolly*. The ram was of quite another breed, being (besides having other marks of difference) of a *chocolate* colour, and *hairy* like the goat. The progeny were of course crosses, bearing, however, a great resemblance to the male parent.

The next season Dr. Wells procured another ram of precisely the same breed as his ewes. The progeny of this second connection showed distinct marks of resemblance to the former ram in *colour* and *covering*. And the like phenomenon, occurring under the like circumstances, was noticed in the lambs of some other adjoining estates in Grenada, and was the occasion of equal surprise and perplexity to the owners of the animals.

2. Six very superior pure-bred *black*-faced *horned* ewes, the property of Mr. Harry Shaw, in the parish of Leochel Cushnie, in Aberdeenshire, were tupped in the autumn of 1844, some of them by a Leicester, *i.e.*, a *white*-faced and *polled* ram ; others of them by a Southdown, *i.e.*, a *dun*-faced and *polled* ram. The lambs thus begotten were crosses.

In the autumn of 1845 the same ewes were tupped by a very fine pure blacked-faced horned ram, *i.e.*, one of exactly the same breed as the ewes themselves. To Mr. Shaw's astonishment, the lambs were all without exception *polled* and *brownish* in the face, instead of being black-faced and horned.

In autumn 1846 the ewes were again served with another very superior ram of their own breed. Again the lambs were mongrels. They did not, indeed, exhibit so much of the characters of the Leicester and Southdown breeds as did the lambs of the previous year ; but two of them were *polled*, one *dun*-faced, with *very small* horns, and the other three, *white* faced, with *small round horns only*. Mr. Shaw at length parted with those fine ewes, without obtaining from them one pure-bred lamb.

(2.) Of the analogous phenomenon, in any of the aspects formerly specified, in the human species, I have procured no additional examples. Inquiries are on foot, in one or two of the West India Islands, as to whether the offspring of negro parents, after the female has had children by the European, ever exhibit traces of the latter ; but the results have not yet reached me. Dr. Maunsell, of Dublin, however, has mentioned to me a case, bearing on the communication of secondary syphilis,—in too imperfect a form, indeed, to warrant its being adduced as an instance of the kind in question, but valuable as

justifying the supposition formerly advanced, that
a woman may somehow acquire syphilis, but have
it in the *latent* form, and subsequently give proof
of the reality of the fact by the birth of a syphilitic child, got by a perfectly healthy man. It is
the case of a woman, who, as far as Dr. Maunsell
could learn, never herself exhibited any signs of
syphilis, yet produced a syphilitic child in a second
marriage, with a husband who never had the
disease.*

With regard to the communication of secondary
syphilis, in relation to Mr. M'Gillivray's theory,
Mr. (Sir James) Paget, of St. Bartholomew's Hospital, makes an important suggestion. "I would
venture to suggest," Mr. Paget writes me, "that
you should try to find whether ever a woman derives
secondary syphilis from her husband, *unless she con-*

---

* Since the publication of my former paper I find that Dr.
Montgomery of Dublin has been beforehand with me in this question as to syphilis ; and that he seems virtually, though obscurely,
to enunciate the doctrine of the *constitutional* character of the
phenomenon exemplified in Lord Morton's mare and Mr. Western's
breed of pigs. Referring to these well-known cases, Dr. Montgomery remarks :—"Such occurrences appear forcibly to suggest
a question, the correct solution of which would be of immense importance in the history and treatment of disease. Is it possible
that a morbid taint, such as that of syphilis, for instance, having
been once communicated to the system of the female [by a con-
ception], may influence several ova, and so continue to manifest
itself in the offspring of subsequent conceptions, when impregnation has been effected by a perfectly healthy man, and the system
of the mother appearing to be at the time, and for a considerable
period previously, quite free from the disease? My belief is certainly in favour of the affirmative."—*Exposition of the Signs and
Symptoms of Pregnancy,* p. 18.

ceives by him. Facts bearing on this point might prove that secondary syphilis is not communicated directly by the seminal fluid, but by the child begotten with it; and this mode of inoculation being proved would go far to prove the foundation of your [Mr. M'Gillivray's] theory." I fear it will turn out, on inquiry, that secondary syphilis may be transmitted directly by the seminal fluid, independently of conception; but perhaps it may appear also that its transmission in this way is *occasional* only and *uncertain*, while it is *very frequent*, or almost *inevitable*, when conception follows intercourse. And a comparative observation of this kind, if clear and undoubted, would be nearly equally decisive.*

(3.) In my former paper, I represent Mr. M'Gillivray as holding that there is a direct vascular

---

* Important as is Mr. Paget's suggestion, it may be difficult successfully to follow it out. It appears, indeed, that of the children born syphilitic in the middle and higher classes of society, a very large proportion derive the virus from the father—a circumstance in itself favourable for the prosecution of the inquiry (see Brit. and For. Med. Chir. Rev., No. XII., p. 348). But in several such cases the mother never exhibits any manifest indication of the virus in her own person (Op. cit., p. 347—Maunsell and Evanson on Diseases of Children, 5th edit., p. 452) ; and although she may not have imbibed the poison, the case given in the text on the authority of Dr. Maunsell is sufficient to show that the only certain criterion of her immunity may be her bearing a non-syphilitic child in a second marriage with a perfectly healthy husband—a test which can be available only in a very few cases. If, therefore, such cases of latent syphilis in the female are common, Mr. Paget's inquiry may fail of an affirmative result only from inability to test them. Possibly, however, the instances of developed syphilis in the female, consequent on conception, may be numerous enough, and sufficiently decided, to lead to that result.

connection, and a continuous interchange of blood,
between the fœtus in utero and its mother, and
also as regarding that sort of connection to be
essential to the validity of his theory of inoculation. Mr. M'Gillivray, however, as I have since
discovered, does not hold any such opinion, nor
does he rest his theory on· so insecure a basis
as I fancied. His ideas regarding the structure
of the placenta, and the connection between the
mother and fœtus, appear, in fact, to be substan-
tially those suggested by the late Dr. John Reid
and others, and now commonly received.

The mistake committed by me arose from some
expressions dropt by Mr. M'Gillivray in the course
of his paper, taken in connection with a case given
by him, and showing an *empty* state of the blood-
vessels of a fœtal calf, whose mother had died of
a hæmorrhage from the lungs, of three days' continuance. As this case seems to me to afford a
stronger support to his theory than I at first imagined,
and to be besides (though not unique *) a valuable

---

* There is a case, by Mery, in the "Mémoires de l'Académie
Royale des Sciences," 1708. But this case is too briefly recorded
to be of any physiological value whatever. It is as follows :—
"Une femme grosse, qui touchait à son terme, se tue d'une chute
très rude presque sur le champ. On lui trouve 7 à 8 pintes de sang
dans la cavité du ventre, et tous ses vaisseaux sanguins entièrement
épuisés. Son enfant était mort, mais sans aucune apparence de
blessure, et tous ses vaisseaux étaient vides de sang aussi bien que
ceux de la mere. Le corps du placenta était encore attaché à
toute la surface interieure de la matrice, où il n'y avait aucun
sang extravasé."—(Op. cit., p. 37). Very possibly, in this case,
as suggested by my friend Dr. J. M. Duncan, of Edinburgh, the
womb may have been ruptured on its external aspect over the

contribution to our physiological knowledge, I shall
introduce it here, and take occasion to offer some
remarks upon it :—

A cow in calf, and past the ordinary period of utero-gestation,
had, from some unknown cause, a slight rupture of the left lung.
A constant hæmorrhage from the fissure proved fatal in three days.
On a post-mortem examination it was found that the hæmorrhage
had taken place partly into the cavity of the chest, and partly
into the air-passages of the lung. A great part of the blood that
entered the lung had found its way upwards, and, being swallowed
by the animal, was passed along with the egesta. The animal
survived until but a very small quantity of blood remained in
the carcase.

The uterus, with its contents, was removed entire, and very
carefully and minutely examined by Mr. M'Gillivray. All the
vessels of the chorion, amnion, &c., were white, flaccid, and
empty. On making a section of the umbilical cord no blood
followed ; on applying a sliding squeeze on that portion of the cord,
and in a direction from the fœtus, slight traces of blood appeared
at the cut ends of the umbilical arteries, but there was no flow,
not even drops. There was no blood in any part of the aorta or
vena cava ; none in the carotids or jugulars near the head ; none
in the external or internal iliac arteries, and large veins in the
ilio-femoral region ; none in the left ventricle of the heart—the
right containing a small coagulum which might amount to about
half an ounce of blood. In short, there were not three ounces
of blood in the calf, taken with all its membranes and placenta.*

On the perusal, formerly, of this case I made but
little account of it; and being impressed with the
unqualified allegation of various physiologists, that

centre of the placenta, and so deeply as to lay open both the fœtal
and uterine vessels,—in which event the bloodless condition of the
fœtus would be referable, not to the passage of its blood by *ab-*
sorption into the maternal vessels, but to the direct escape of the
blood from its own vessels.

* Abridged from Mr. M'Gillivray's paper in *Aberdeen Journal*
for March 21, 1849.

" when a pregnant animal dies of hæmorrhage, the vessels of the fœtus remain full of blood,"* I too hastily concluded that some fallacy must attach to it, or at least that, however it might be explained, it was too slender a foundation whereon to rest the doctrine (erroneously assumed as that intended) of a direct vascular communication.

Mr. M'Gillivray's real object in adducing the case was to combat the opinion entertained by some physiologists, that, while the fœtus receives supplies from the mother through the placenta, it returns nothing from its own system to hers. " I am quite aware (he says) that many physiologists maintain that, in the highest species of animals, the blood cannot be returned from the fœtus to the mother during utero-gestation." That this opinion is very generally held by physiologists in this country is quite certain. Dr. Alison, for instance, after observing (on the authority of Magendie and of Dr. David Williams, of Liverpool), that camphor and oil injected into the blood of pregnant animals are soon detected in the blood of the fœtus; but that poison, injected into the umbilical arteries, although mixing with the blood on its way from the fœtus to the placenta, does not affect the mother; and that fatal hæmorrhage in the mother does not apparently diminish the fulness of the vessels of the fœtus,— adds, " so that it would seem that the transmission of fluids is almost entirely from the mother to the fœtus."†

---

* Magendie ; Physiology, by Milligan ; 2nd edit., p. 569.   Dr. D. Williams, Edin. Med. and Surg. Journal, Vol. xxv., p. 102.

† Outlines of Physiology, 3rd edit., p. 426.—In his History of

Again, Dr. Kirkes, referring to Professor Goodsir's observations as to the intervention of two distinct layers of nucleated cells between the fœtal and maternal portions of the placenta, speaks of the one being " probably designed to separate from the blood of the parent the materials destined for the blood of the fœtus," while the other " probably serves for the absorption of the material secreted by the other set of cells, and for its conveyance into the blood-vessels of the fœtus," *—no idea, seemingly, being entertained of a converse process. Moreover, the view taken by most physiologists of the destination of that portion of the fœtal blood which is transmitted to the placenta appears to be exclusively that of *renovation* or *aeration*, by coming into relation with the oxygenated blood of the mother ;†—nothing being said as to *re-absorption* into the maternal system.‡

Medicine, Dr. Alison expresses himself even more strongly on the subject :—" The experiments of Magendie and others have proved that any substance which may be circulating in the blood of the mother finds ready access to that of the fœtus, but that there is little or no transference of fluids in the opposite direction."—Cyc. of Pract. Med., Vol. i. p. lxxxiii.

* Handbook of Physiology, p. 643.

† Carpenter, Principles of Human Physiology, 2nd edit., p. 718. Manual of Physiology, p. 474.

‡ It may be asked, whether the idea expressed by the terms "renovation" and "aeration" does not necessarily include that of the transference of *some kind* of matter from the fœtus to the mother ? Supposing "that the umbilical arteries terminate in the umbilical veins, and not in the vessels of the uterus, and that the [whole] blood in the umbilical arteries ' passes from the arteries into the veins, as in other parts of the body, and so back again into the child ' "—(Dr. J. Reid, *Researches*, p. 318)—still this blood

Mr. M'Gillivray brings forward this exsanguine
fœtal calf to depone in opposition to these exclusive
views, and by its testimony to prove that of the blood
which returns to the placenta a *portion* is *absorbed*

---

is believed to have acquired something *effete* in its transit through
the fœtal system. What becomes of this effete fœtal matter ?
Clearly there is no outlet for it but *through* the mother's system,
and by *her* excretory organs.

Judging from the analogy of the process of aeration in the adult,
and from the condition of the fœtus, which renders all excretion
by its own organs (except of bile into the intestine) impossible,
carbonic acid and the elements of urine probably form the chief
part of that effete matter. Perhaps some *animal* matter may attach to it —such as is thrown off with the watery vapour in the
adult, and amounts, according to Collard de Martigny, to about
three parts in 1,000 of the vapour. This animal matter, however, may be thought too inconsiderable in amount, or not of a
nature to exert any influence on the maternal constitution. But
no one, probably, will think so, who reflects on what is now
admitted in regard to blood-diseases, and the small influences by
which the whole mass of blood may be affected.—(See Kirkes'
Physiology, pp. 71, 286, 292.) The "materies morbi" of
scrofula, of syphilis, and of many chronic diseases, must often
exist in the blood in an impalpable form,—nay, must even be
evolved in the secretions and the plastic exudations. More
curious still, even MIND itself, *immaterial* as it is, must pervade
the blood (though its acknowledged seat is the nervous system),
and must impart itself to the semen and the microscopic ovum !
"In the most perfect animals (says Müller), and even in man, we
must suppose that the ovum and semen contain within themselves
all the conditions necessary for the production of a new being
endowed with *mind;* and, consequently, that one or both of them
contain the *essence* of mind in a *latent* state."—(*Physiology,* Vol.
i., p. 820.)

Is it more extraordinary, or at all fanciful, to suppose, that a
subtile *materies fœtus* may attach to the effete matters which pass
into the blood of the mother, and that it may *inoculate* her
system with its own *distinctive* qualities ?

by the uterine veins, and re-enters the circulation of the mother. That it furnishes " undeniable " evidence of this (as he alleges) is more than can fairly be affirmed, because it is quite conceivable that, without giving back to the mother a single drop of its blood, the fœtus may, during the three days, have simply used up the blood within it. Whether this happened it is difficult to say. But it may be questioned whether the vital actions of the fœtus would not have failed long before its blood was actually expended to this extent ; and it seems altogether much more probable that the nearly complete disappearance of that fluid was owing *in part* to its being abstracted by the mother. That, under the circumstances in which the mother's system was placed during the continuance of the hæmorrhage, a powerfully *derivative* effect had been produced on the general mass of her blood, withdrawing a large part of it from the several organs (including the maternal portion of the placenta) towards the seat of the hæmorrhage, seems very certain from the observations of Haller and Spallanzani; and that her general function of absorption had been unusually active, the well-known experiments of Magendie sufficiently demonstrate. And we should scarcely conceive, *à priori*, either that the blood of the fœtus had not been brought within the sphere of action of those two influences, and, unless an obstacle interposed, been laid under contribution for the support of the mother's system ; or that the double layer of nucleated cells, intervening between the maternal and fœtal vessels, and which habitually allows the

E

passage of fluids from the mother to the fœtus, had then offered any impediment to their passage in the opposite direction.

The discrepancy between Mr. M'Gillivray's case and the statements of Magendie and Williams, as to the *fulness* of the fœtal vessels when a pregnant animal dies of hæmorrhage, is probably only an apparent one. Those statements have reference to cases where the mother dies *suddenly* from *profuse* and *rapid* loss of blood, and where the emptying of the fœtal vessels in the only way in which they can be drained, viz., by absorption, is *anticipated by death;* while in Mr. M'Gillivray's case the hæmorrhage being comparatively gradual, *time was given* for the exercise of that process, and circumstances concurred to render it *unusually active.*

In the absence of any *experimentum crucis*, whereby a positive solution of the question might be had, the anæmic condition of the fœtal calf may, I think, with great probability be ascribed, in part, to the expenditure of its blood in the nutritive processes going on within it, and in part, as Mr. M'Gillivray supposes, to the re-absorption at least of its more fluid constituents into the system of the mother. And what seems to have been thus possible, and, under the circumstances, may very fairly be presumed to have actually occurred, and to have taken place pretty rapidly, may reasonably be held to imply, that, at all times, though ordinarily but slowly, there goes on within the gravid uterus a gradual removal by absorption, and reception into the maternal system, of such portions of the fœtal blood as are unfitted by reason of impurity for

ministering to the nourishment of the fœtus, and
for the *renovation* of which the placenta may be
inadequate. It is not probable, indeed, that the
amount returned is very great, and the experiments
of Magendie are sufficient to demonstrate that it can
be but small. If, however, it take place at all, be
the quantity absorbed within a given time never so
inconsiderable, it is all that is necessary for the
support of Mr. M'Gillivray's theory,—the essential
principle of which cannot, perhaps, be better expressed than in the words applied to signify the
influence of moral causes on human character—" a
little leaven leaveneth the whole lump."

Wagner, indeed, expresses a decided opinion that
a *mutual* interchange of fluids takes place in the
placenta,—" *the blood of the mother abstracting matter
from that of the fœtus*, and the blood of the fœtus
taking in its turn matter from that of the mother."*
Müller, also, is of the same opinion. "In mammalia, the vascular villi of the fœtus are received
into the vascular sheaths of the uterine placenta, so
that the capillaries of the fœtal and those of the
maternal system come into contact with each other,
*and suffer an interchange of the matters which they
contain.*"† Neither Müller nor Wagner, however,

---

* Elements of Physiology, translated by Willis, p. 202.

† Elements of Physiology, translated by Baly, Vol. ii. pp.
1604–5. Dr. John Reid makes an observation which seems to
convey the very same idea:—"The blood of the mother contained
in the placental sac, and the blood of the fœtus centained in the
umbilical vessels, can readily *act* and *react* upon *each other.*"
—*Physiological, Anatomical and Pathological Researches*, p. 326.

refer to any facts in support of their opinion as
regards the absorbent action of the mother on the
fœtus.   Probably they rest it on analogical grounds.
And, indeed, all that seems necessary to give confirmation to it is proof that poisons introduced into
the umbilical arteries, or into the body of the fœtus,
may pass into and affect the mother.   That
Magendie's experiments failed to show this is well
known ; but, valuable as they may be allowed to be,
they do not demonstrate that such absorption may
not take place—still less, that it is impossible.   The
only negative inference that can be drawn from them
is, that within the time occupied by them, and
under the circumstances in which they were per-
formed, no obvious absorption occurred.   And it is
conceivable that they may have been too few in
number, or too little varied—nay, even too unskil-
fully conducted to be of much value, and may
demand repetition ; or,   that the circumstances
under which experiments of this kind must necessarily be made are such as to preclude the hope of
our ever obtaining satisfactory information through
this channel.   Yet, even in this event, we may yet
obtain the requisite satisfaction.   Possibly, nature
may herself yet furnish us, in her own way, with
the evidence which we cannot extract from her by
artificial methods of interrogation.   Had we, *e.g.*,
undoubted instances of the fœtus being tainted with
syphilis through the father, and communicating the
virus to the mother, we should have all the proof
we need desire that matters can and do pass by absorption from the fœtus to the mother.   Whether

such evidence shall yet be had remains to be seen ; but it may be observed that Mr. (Sir James) Paget has suggested the criterion by which the value of alleged instances of this kind may be determined.

ABERDEEN, *August* 18, 1850.

ESSAY THIRD.

ON THE FŒTUS IN UTERO
AS INFLUENCED IN ITS DEVELOPMENT
BY
MENTAL STATES IN EITHER PARENT.

FROM
THE EDINBURGH MONTHLY JOURNAL OF MEDICAL SCIENCE
For *November*, 1850.

ESSAY THIRD.

ON THE FŒTUS IN UTERO, Etc.

THAT mental causes, or states of mind, operating within the female during pregnancy, or within either parent at the time of coitus, may variously influence the nutrition and development of the fœtus, has long been matter of popular belief; and, setting a goodly number of recorded instances to the account of old wives' fables, this belief may be allowed to have a stable foundation in facts.* It is not my intention, however, to enter on the general subject further than as it bears on Mr. M'Gillivray's theory, referring merely to such facts as serve to show that mental causes may so influence the growth of the fœtus as to produce results analogous to those ascribed to inoculation by a former fœtus,

* Any one curious in cases of this kind may advantageously consult Dr. Allen Thomson's article on "Generation," in the Cyc. of Anat. and Physiology ; and also an article in the Edin. Med. and Surg. Journal, Vol. xxv., p. 134.

and therefore to exhibit a source of fallacy in the reference of these to such inoculation as their cause.

(1.) Of the cases of this kind now to be noticed some involve changes both in the configuration and the colour of the progeny—others, as far as appears, changes in the colour only or chiefly.

In Daniel's " Rural Sports " the following details are given respecting the setter-bitch and cur-dog formerly referred to :—

" As the late Dr. Hugh Smith was travelling from Midhurst into Hampshire, the dogs, as usual in country places, ran out barking as he was passing through the village, and amongst them he observed a little ugly cur, that was particularly eager to ingratiate himself with a setter-bitch that accompanied him. Whilst stopping to water his horse, the doctor remarked how amorous the cur was, and how courteous the setter seemed to her admirer. Provoked to see a creature of Dido's high blood so obsequious to such mean addresses, the doctor drew one of his pistols, and shot the cur. He then had the bitch carried on horseback for several miles. From that day the setter lost her appetite, ate little or nothing, had no inclination to go abroad with her master, or to attend his call; but seemed to pine like a creature in love, and express sensible concern at the loss of her gallant. Partridge season came, but Dido had no nose. Sometime after she was coupled with a setter of great excellence, which, with no small difficulty, had been procured to have a breed from, and all the caution that even the doctor himself could take was strongly exerted, that the whelps might be pure and unmixed. Yet not a puppy did Dido bring forth but was the *exact picture* and *colour* of the cur that had so many months before been destroyed. The doctor fumed, and, had he not personally paid such attention to preserve the intercourse uncontaminated, would have suspected that some negligence had occasioned his disappointment ; but his views were in many subsequent litters also defeated, for Dido

never produced a whelp which was not exactly similar to the un-
fortunate cur who was her first and murdered lover."*

In Mr. Blaine's " Encyclopædia of Rural Sports "
this other case is given :—

" The late Lord Rivers [says Mr. Blaine] was famed for a breed
of black and white *spaniels*, one of which, having more than the
usual quantity of white, he presented to us. We had, at the same
time, a *pug-bitch* of great beauty. The attachment of this bitch
to the spaniel was singularly strong. When it became necessary
to separate her, on account of her heat, from this dog, and to con-
fine her with one of her own kind, she pined excessively ; and,
notwithstanding her situation, it was some time before she would
admit the attentions of the pug-dog placed with her. At length,
however, she was warded by him, impregnation followed, and at
the usual period she brought forth five pug-puppies, *one of which
was perfectly white, and rather more slender* than the others,
though a genuine pug. The spaniel was soon afterwards given
away. At two subsequent litters (which were all she afterwards
had) this bitch also brought forth a *white* pug-pup, which the
fanciers know to be a very rare occurrence. It is also a curious
fact that each succeeding white puppy was *less slender* in form
than the preceding, though all were equally white."†

The two cases now given have many points in
common, and appear to be free from any material
source of fallacy. In the former there was not even
sexual intercourse—much less fruitful intercourse—
between the bitch and the cur, to whom her progeny
bore so decided a resemblance ; and in the latter, if
intercourse occurred, which it appears did not, there

---

* Daniel's Rural Sports, Vol. iii. pp. 333, 334.—A case very
similar to the above, occurring in a bitch belonging to him, has
been mentioned to me by Mr. Walker, Portlethen, in Kincardine-
shire.

† Encyclopædia of Rural Sports, by Delabere P. Blaine, Esq.,
p. 412.

was no result from it. In both females there seems
to have been a strong and abiding attachment to
the cur and spaniel respectively, and an equally keen
and enduring emotion of disappointment at being
separated from them. The resemblance of the
progeny, or of certain of these, to the dogs, appears
to have been of too special a kind to admit the supposition of its being accidental. There need be no
question, therefore, but that the cases are fair examples of mental impressions of a permanent
character so operating on the female parents as to
influence the development of their offspring while
in utero, and that in such a manner as to cause them
to resemble the male animals that were the objective
causes of those impressions.

Nearly similar remarks apply to the following
case, for the particulars of which I am indebted to
Dr. John R. Trail, of Monymusk, in Aberdeenshire.
The chief point of interest in this case lies in the
resemblance being at once general, and yet extending to an abnormal peculiarity of conformation, and
in its thus forming a connecting link between the
two foregoing cases and two others presently to be
adduced. It does not appear, indeed, whether the
mother manifested any special attachment to the
male animal which her offspring resembled, but
such attachment may reasonably be presumed :—

A mare and a horse (a gelding), belonging to a friend of Dr.
Trail, had for some years worked together on the same farm, occupied adjacent stalls in the same stable, and pastured together in
summer in the same fields. The gelding was of a black colour,
with white legs and face, and had a singular peculiarity in the

form of the hind legs, which, when the animal was standing, appeared as if quite straight, there being no appearance of the leg being bent at the hough-joint, as in ordinary cases ; the pasterns likewise were very long, so as to cause the feet to look as if placed almost at right angles to the legs.

Having been some years thus associated with this gelding, the mare was covered by a stallion of the same colour with herself— both stallion and mare being of a bay colour, with black legs, and a small spot of white only on the forehead. The foal which was the produce of this connection very exactly resembled the gelding in *colour*, and in the *shape*, too, more particularly of the hind legs, as above described. " From the description I have attempted to give you [Dr. Trail writes me], you could not form any very distinct idea of the peculiar conformation of the horse ; but the resemblance of the foal to him was remarkably clear."

Dr. Montgomery, of Dublin, gives the following interesting case :—

" A lady, pregnant for the first time, to whom I recommended frequent exercise in the open air, declined going out as often as was thought necessary, assigning as her reason, that she was afraid of a man, whose appearance had greatly shocked and disgusted her ; he used to crawl along the flag-way, on his hands and knees, with his feet turned up behind him, which latter were malformed and imperfect, appearing as if they had been cut off at the instep, and he exhibited them thus, and uncovered, in order to excite commiseration. I afterwards attended this lady in her lying-in ; and her child, which was born a month before its time, and lived but a few minutes, although in every other respect perfect, *had the feet malformed and defective, precisely in the same way as those of the cripple* who had alarmed her, and whom I had often seen."*

Nothing can be more appropriate than Dr. Montgomery's short commentary on this case :—" Here was an obvious and recognized object making a powerful impression, of a disagreeable kind, com-

---

* Exposition of the Signs and Symptoms of Pregnancy, &c., pp. 16, 17.

plained of at the time, and followed by an effect in
perfect accordance with the previous cause, there
being between the two a similarity so perfect, that,
with Morgagni, I ' will not easily suppose that chance
could have been so ingenious, if I may be allowed to
speak thus, and so exact an imitator.' "

A case very similar to the foregoing has been
communicated to me by a distinguished English
physiologist—himself personally cognisant of the
facts.   It is as follows :—

" A lady, when two or three months pregnant, was accosted by
a *one-armed* beggar, who, on her refusing to relieve him, menaced
her violently, so as to alarm her seriously, and shook his stump
at her. She was extremely agitated, and during the whole re-
mainder of her pregnancy was under the firm expectation that her
child would be one-armed—which was the case. This child is now
grown up to manhood, and occupies a highly respectable position
in society."

In this case (and the remark applies equally to
Dr. Montgomery's) the effect was *partial* only, and
on this account it may be thought not altogether
pertinent to the subject under consideration; but it
probably embraced all that was peculiar in the object,
as it certainly did all that was striking to the mind
of the lady.   And had the occasion of the impression been some well-marked but normal peculiarity
of the features—*e.g.*, of the nose, instead of a
deformity of the arm; and had the man been the
object formerly of the lady's love and attachment,
and still during her pregnancy of her cherished
regard, and had the resemblance in her child extended to the part supposed, the effect, though

equally partial, might have seemed more general. The case might thus have been set down as a fair example of the power of the imagination, in a pregnant female, to cause her offspring very exactly to resemble an individual not its father. Of this the following is, perhaps, an instance :—

" A young married woman, residing in Aberdeen, between whom and a young man a strong attachment and a matrimonial engagement had long existed, but who were never married, and never had sexual intercourse together, gave birth to a child, which bore so striking a resemblance in its features to the woman's first lover as to attract the notice of herself and many others of the acquaintance of the parties."

In this case—communicated to me by Mr. Robert M. Erskine, surgeon, here, who was well acquainted with the individuals concerned, and had personally satisfied himself of the accuracy of the fact—the resemblance may have existed only in the imagination of the observers, or been magnified through the love of the marvellous,—and, giving it as given to myself, I adduce it merely as a possible example of what may be a real occurrence, and would contrast remarkably with the observation alluded to by Dr. Allen Thomson,—viz., "that the human female, when twice married, bears occasionally to the second husband children resembling the first, both in bodily structure and mental powers,"—and also with the cases given as instances of this, in the Postscript to Essay First.

It appears that many breeders of stock are impressed with the belief, that certain *colours* present

to the eye of the parent animals, and particularly
of the female, at the time and in the act of their
being coupled together,—and to the eye of the female
both before and during her pregnancy, influence the
colour of the progeny; and that they make this
belief a practical principle of action in the breeding
of their stock, in order either to prevent or to secure
the admixture of any particular colour in the offspring different from that of the parent animals.
" We know," says an anonymous writer, " a great
breeder of pure Angus stock [black polled breed],
who makes it a rule to have every animal about his
farm of a black colour, down to the very poultry."*
And an eminent breeder of the same stock in this
county informs me that he extends this rule to the
steadings in which his cattle are kept.

To illustrate generally the grounds of this belief
and practice the following cases may be cited :—

(*a*) A black polled [Angus] cow, belonging to Mr. Mustard, a
farmer in Forfarshire, came into season while pasturing in a field
bounded by that of a neighbouring farmer. Out of this last there
jumped into the other field an *ox*, of a *white colour*, with black
spots, and *horned*, which went with the cow till she was brought
to the bull,—an animal of the same colour and breed as herself.
Mr. Mustard had not a horned animal in his possession, nor any
with the least white on it : and yet the produce of this (black and
polled) cow and bull was a black-and-*white* calf, with *horns*.†

(*b*) Last year (1849) twenty cows of the black polled Angus
breed,—belonging to Mr. William M'Combie, in this county, and

---

* *North of Scotland Gazette* newspaper for July 17, 1849.
† Library of Society of Useful Knowledge, volume on Cattle,
p. 171.—This seems to be the same case as that given by Dr.
Allen Thomson, in Cyc. Anat. and Phys., art. " Generation,"
Vol. ii. p. 474.

whose stock is perhaps the finest in the kingdom,—produced as many calves, all of them black and polled, except one single calf, which was *yellow*-and-*white spotted.* Mr. McCombie had, as usual with him, taken the precaution of causing the cows, both before and during their pregnancy, to mix with none save perfectly black cattle, except in respect of the mother of this calf. This cow had unwittingly been put to an out-farm, to be starved, in order to fit her for the bull. There, for a considerable period prior to her being served with the bull, she had grazed with a large *yellow*-and-*white spotted ox:* of this ox the calf she subsequently bore was the very picture,—the likeness, however, extending no farther than to the colour, and the calf still retaining the shape and configuration of its parents, which were both of the same breed and colour.

(c) Out of a large herd of cows, of the pure Teeswater breed, all of them of the brown or roan colour (belonging to Mr. Cruickshank, Sittyton, near Aberdeen), there is every year dropt one, or at most two, *white* calves : these, in order to prevent the introduction of this colour among the cattle, are invariably sold, and sent away. Last year, however, concurrently with the *whitewashing* of all the farm-steadings, the very large number of twelve white calves were produced. And the like occurrence happened last year also, in the herd of an extensive breeder of the same kind of stock, in Yorkshire, in connection with the like process of whitewashing,—this process having, in both cases, been very extensively carried out before the breeding season began, with the view of preventing the breaking out of the pleuro-pneumonia, then epidemic in the neighbourhood, and very destructive.*

(d) "At the time when a stallion was about to cover a mare, the stallion's pale colour was objected to, whereupon the groom, knowing the effect of colour upon horses' imaginations, presented before the stallion a mare, of a pleasing colour, which had the desired effect of determining a dark colour in the offspring. This is said to have been repeated with success in the same horse more than once."†

* Communicated by Mr. Cruickshank, who informs me further, that he has had too many proofs of the agency of the cause in question to allow him entertaining any doubt on the subject.

† Art. "Generation," in Cyc. Anat. and Phys., vol. ii., p. 474.

(e) " I was told (Mr. McCombie writes me) by an old servant of mine, Morrice Smith, that, when he was a servant in the parish of Glass (Aberdeenshire), a black bull served a black cow at the time when a white mare passed them, and that the produce was twin *white* calves. There were no white cattle upon the farm where this occurrence happened.' "*

Such cases as several of those now cited can scarcely fail to recall to the reader's mind the story given in the book of Genesis (chap. xxx.), of Jacob and his peeled rods, and the effect of these in causing the flocks, before whom they were placed at the time of conception, to bring forth ring-straked, speckled, and spotted cattle.

It does not appear from the sacred narrative whether the influence of the rods was exerted on the minds both of the male and female cattle, or confined to those of the female. But it seems probable (Gen. xxxi. 5-12) that the effect was a *supernatural* one, and designed to enrich the needy patriarch at the expense of the crafty Laban, by whom, for fourteen years, he had been sore let and hindered " in providing for his own house ;" and that Jacob

---

* My friend Dr. J. M. Duncan writes me, that he has " more than once heard farm-servants say, that it is a sure plan to get a white foal, to hang up a pure white sheet before the mare when she conceives." Probably hanging up such a sheet in the stable during the whole period of pregnancy would be equally effectual. And another correspondent (J. Warwick, Esq., Sussex House, Hammersmith) says : " The belief in the effect of external objects on the imagination of pregnant women is so strong in Italy, that females in that condition keep in their rooms, and before their eyes, small wax figures of *a pretty child.* This is called 'Il Bambino,' represents the infant *Jesus,* and is especially worshipped and adored, with the view of procuring beautiful offspring."

had beforehand, in a dream, intimation given him of the design, as well as a sensible representation both of the agency to be employed and of the result which was to follow.* If this were the character of the transaction, it would be unwarrantable to draw any inference from it in relation to the present inquiry, unless it could be shown, which I apprehend it cannot, that the same expedient will now, as then, produce the like results. At the same time it may, perhaps, be not unfairly referred to, in illustration of ordinary phenomena of a somewhat analogous kind. Read by the light of these, it may be held to indicate that the Almighty accomplished His purpose simply by *enhancing* a natural agency—" moulding it secretly in the hollow of His hand,"—and so

---

* Dr. Allen Thomson, in his comment on this transaction (Cyc. Anat. and Phys., art. " Generation "), seems to me to have very completely missed the import of the essential parts of the narrative. He gathers from it, that Jacob had parti-coloured males to breed with, and makes the result an affair of mere hereditary transmission. But Jacob had not this advantage, and it seems clearly to have been part of his own proposal to Laban that he should not. Laban, at least, on the bargain being concluded, very carefully singled out all the animals of that sort—males as well as females —even " to every one that had some white on it, and all the brown among the sheep ; " and, giving them into the hand of his sons, set three days' journey betwixt this parti-coloured flock and that which was to be tended by Jacob. A difficulty, it may be added, attaches to the right understanding of the whole matter, from the obscurity of v. 40 (chap. xxx.)—the true meaning, if not the proper rendering, of which appears to be this :—" And Jacob did separate the lambs, and set aside from the flock all the ring-straked and all the brown in the flock of Laban ; and set them apart for a flock to himself, and put them not unto Laban's cattle."

be regarded as reflecting back light in its turn upon that agency. The validity, however, of this appeal must turn upon our being able to satisfy ourselves, on independent grounds, that there is in nature an agency of this kind; while the precise value of the appeal will hinge on the extent to which we can thus ascertain that it ordinarily operates,—questions these for the solution of which we have not at present the requisite data.

(2.) All that need be said in the way of direct inference from the facts brought together in this division of the subject, may be comprised within a narrow compass.

Supposing the statements respecting them to be authentic—and no question, I apprehend, as to this can well be raised—the cases are nearly unequivocal. The only fallacy that can attach to them, is that arising from the possibility, that the peculiarities in the progeny were either purely *accidental*, or owing to corresponding qualities *latent* in the parents, but breaking out in the offspring. The relation, however, in most of the cases between the peculiarities in question and their presumed causes, is too close and of too special a character to admit of either supposition. We are, therefore, well entitled, I think, to regard the greater number, if not the whole of them, as examples of mental causes so operating either on the mind of the female, and so acting on her reproductive powers, or on the mind of the male parent and so influencing the qualities of his semen, as to modify the nutrition and development of the offspring.

How, in respect of the female, this influence is exerted,—what its *modus operandi,*—and what the conditions of its agency, it is not easy to determine. The mental affections seem to have been in most of the cases, and were probably in all of them, of a *strong* and *enduring* kind ; and we can easily conceive this to have been essential to the result. That the alteration in the growth of the fœtus was determined solely, as is vulgarly supposed, by the *images* in the mind of the mother,—*i.e.,* by the mere *sensations* and *perceptions* therein produced, independently of the *emotions* excited by them,—cannot well be supposed. It is, doubtless, to this "*compound state*" of mind— to use an expression, without adopting, however, the psychological theory, of Sir James Mackintosh— " easily called to mind," in consequence of the vividness of its first impression, " frequently recurring," and "warmly felt,"* that we must ascribe the effect.

It is not unlikely that this particular agency of the mind is more frequently exerted in the females of the lower animals than in those of our own species ; and that cases exemplifying it are oftener met with in the brute than in man. If this be so, a reasonable explanation of the fact may be given. We know that the minds of the lower animals are in a great measure limited to *particulars,* and these few in number, and almost exclusively external objects of sense ; that the external senses are more perfect in them than in us ; and that the perceptions result-

* Ethical Philosophy, edited by Whewell, 2nd edition, pp. 397, 398.

ing from their exercise seem, in various instances,
to follow more surely and more quickly—to be
more intuitive and wider in their scope, and more
vivid—in them than in man ; and that the simpler
emotions (excited by those perceptions) of joy, fear,
affection, anger, &c., of which they are manifestly
susceptible, seem often to be peculiarly strong.*
We know also that they possess the faculty of
memory ; and we may well suppose, from their
limited range of association (or suggestion), that
sensations that formerly made a powerful impression on their minds, will be more easily and oftener
recalled in them than in us, who, though more apt
to be "troubled about many things," are proportion-
ally less apt to be affected, or at least permanently
or continuously impressed, by any one thing. These
circumstances and peculiarities of mental action
must obviously be singularly favourable to the production of the results in question,—keeping in mind,
that the mental agency under consideration is manifestly closely connected with *sensations* and with
simple *perceptions* and *emotions* thence resulting, and
with a certain *intensity* and *endurance* of all these.

That the parts of the female system specially
affected by this " compound state " of mind are the
blood and nutritive processes generally seems very
probable ; that the *kind* of change, however, therein
produced, is rather *dynamical* than organic, has
been suggested to me by Dr. Carpenter; that it may
be of identically the same nature with the change

---

* See art. "Instinct," in Cyc. of Anat. and Phys.

produced by the inoculation principle is quite possible, the two, however different in their origin, thus running up into, and meeting in, one *common* principle,—to wit, a modification of the constitutional or dynamical powers of the female ; and that it may be more or less *persistent*, and operate on a subsequent conception, independently of a renewal of the mental state itself, some of the cases seem to indicate. That the change in question is produced through the medium of the nervous system, we cannot doubt—"that being the acknowledged seat and instrument of mental acts ; " and that the parts of the nervous system more immediately concerned, in addition to certain great nervous centres (the sensory and emotional ganglia*), are the ganglionic nerves, may reasonably be conjectured—these nerves being the only channels by which the blood and the nutritive processes can be brought into relation with the mind.†

Beyond this we cannot venture to go in the way of speculation. The singular influence thus exerted by the mind of the mother on the growth of the fœtus, is not one " for which," as has been remarked of other modes of action of the mind on the body, "it is likely that we shall ever be able to assign a reason, or which it would be any great hardship to

---

* See Carpenter's Manual of Physiology, p. 522, et seq. ; and Todd and Bowman's Physiological Anatomy, &c., vol. i., chap. xi.

† Alison—Outlines of Physiology, 3rd. edition, pp. 402-4 ; and Physiological Inferences from the Study of the Nerves of the Eye-ball, in " Medical Gazette," No. 705 (June 1841) p. 410, et seq.

be obliged to regard as an ultimate fact in physiology."*

(3.) Reverting now to the theory of inoculation,
and to the cases cited as instances of it, it will be
obvious, from the facts brought together under the
preceding head, that in any ordinary case where an
animal resembles a male, not its progenitor, by
which its mother had on a former occasion been
impregnated, the resemblance may be explained as
well on the principle of mental agency as on that of
inoculation; or, at least, that in ascribing it to this
latter cause a manifest source of fallacy attaches to
the assumption.

To obviate this, and to determine conclusively
whether or not the phenomenon is independent of
mental agency, all that seems necessary is, to institute experiments in breeding on a sufficiently large
number of different kinds of animals, selecting animals of acknowledged purity of blood, and conduct-
ing the experiments so as to exclude (which I understand could easily be done) the agency of any such
mental impressions on the minds of the animals concerned as could reasonably be supposed to influence
the results. Should these be of an affirmative kind,
and be, at the same time, sufficiently numerous and
decided, it would follow conclusively that the phenomenon is not due to mental causes. In connection,
however, with this set of experiments, it would be
desirable to institute another set in which every

---

* Alison—On the Physiological Principle of Sympathy,—
Transactions of Edinburgh Medico-Chirurgical Society, vol. ii.,
pp. 223, 224.

advantage should be given to the agency of mental causes. Should the results from these be equally affirmative—although this circumstance would not vitiate the inference fairly deducible from those of the other set, unless by exhibiting in these any source of fallacy otherwise perhaps imperceptible— they would afford points of comparison and contrast, and thereby enable us the better to appreciate the value of that inference; while, if they should be wholly or in a great measure negative, it is clear that they would add greatly to the conclusiveness of the inference in question.

The following is a detailed outline of the most important of these

EXPERIMENTS.

First Set.—To determine how far the circumstance of a previous impregnation by a male of a different species, breed, or colour, from the female—the agency of mental causes on the result being excluded —influences the conformation or colour of the offspring subsequently borne by her, to a male of the same breed and colour as her own :—

The female to be served the first season by a male of a different species, breed, or colour from her own ; and the following season, by a male of the same breed and colour as herself,—taking these precautions against the agency of mental causes, *first*, and generally, that the female shall not at any time, up to the period of her second *accouchement*, have seen the former of those two males, nor any animal of that species, breed, or colour ; and, *secondly*, and particularly, that the coitus between her and that male shall be so managed as to prevent her from seeing him ; that, subsequently, during this her first pregnancy, the female shall be kept exclusively with animals of her own breed and colour ; and that, immediately on her delivery, the (cross) animal produced be taken from her, and ever after kept out of her sight.

Second Set.—To determine how far mental causes, operating through the external senses, influence the colour and conformation of offspring :—

I. Mental causes operating at the time of coitus only, or at least originating then—

The male and the female, both of the same breed and colour, in being coupled together, to have fully in their view animals of a different species, breed, or colour, or objects of any kind of a different colour, regard being had, in respect of the matter of colour, that this present a marked contrast to that of the animals concerned, and, particularly, that it be of a bright or striking hue.

To prevent, as far as possible, any influence of this kind from operating on the mind of the female during her pregnancy, the female be made to herd exclusively, till after her delivery, with animals of the same breed and colour with herself.

II. Mental causes operating on the mind of the female during pregnancy only—

The male and the female, both of the same breed and colour as before, to be coupled together in the usual manner (or in the presence of animals of the same sort only) ; but the female, during her pregnancy, to herd exclusively with animals of a different species, breed, or colour.

III. Mental causes operating both at the time of coitus, and subsequently during the pregnancy of the female—

The male and the female, in being coupled together, to have fully in their eye animals of a different species, breed, or colour ; and the female, during her pregnancy, to herd exclusively with animals of this sort.

The details now given embrace the principles on which, as it seems to me, decisive experiments might be made, and also the precautions necessary to be attended to, in order, as far as possible, to avoid fallacy in the results obtained from them. They might be very variously modified, and ought certainly to be tried on many different species of animals, in order to give at once a breadth and precision to the results.

Should it be clearly established by such experiments that an animal, the produce of a male and female of the same breed, born subsequently to the impregnation of the female at a former period by a male of a different breed or species from her own, has, from the single circumstance of such previous impregnation of its mother, its nutrition and development so far changed as to resemble that male animal, not its progenitor, the experiments would farther supply materials towards determining the extent to which the fact holds, and the conditions essential to its occurrence. And, in particular, they would furnish more or less satisfactory answers to the following questions,—themselves suggestive, both of modifications in the experiments, and of the points to be attended to in noting the results—

1. Does the fact hold *universally,* or only *pretty often,* or only *occasionally* and *rarely ?*

2. Does the animal so produced partake *decidedly* or only *partially* of the characters of the male animal which it resembles,— in *colour* merely, or in *conformation* also ?

3. Does the degree of resemblance depend on the number of times the mother may have conceived by that male, before bearing the animal in question ?

4. Is the resemblance lessened or in any way affected by the

female having been *first* impregnated by a male of her own breed
and colour, prior to her having borne offspring by that other
male ?

5. Does the resemblance to that male become less and less in
each animal successively produced, and does it ultimately become
imperceptible ?

Whether such experiments as are here suggested
would, if instituted, establish as a principle in the
physiology of generation, that, independently of
mental agency, a male animal, that has once had
fruitful intercourse with a female, may influence
the products of subsequent conceptions of that female
in which he himself is not concerned, it were need-
less at present to speculate.   But the cases given in
this and in Essays First and Second, as instances of
it, taken along with the belief of the frequent
occurrence of such cases, and the acknowledged
importance of maintaining the purity of breed of
most kinds of our domestic animals, may well engage
the attention of the agricultural body, and lead them
to investigate the subject.   How much the Royal
Agricultural Societies of England and of Ireland,
and the Highland and Agricultural Society of Scotland, have it in their power to promote such an
investigation; and how much the prosecution of it
would be in keeping with the ends of their institu-
tion, it is unnecessary to say.

ABERDEEN, September 10, 1850.

ESSAY FOURTH.

ON THE FŒTUS IN UTERO AS THE CHANNEL OF TRANS-
MISSION OF CONSTITUTIONAL SYPHILIS FROM THE
MALE TO THE FEMALE PARENT.

FROM

THE GLASGOW MEDICAL JOURNAL,

For *January*, 1859.

READ BEFORE

THE MEDICAL SOCIETY OF SOUTHAMPTON,

December 2, 1856.

ESSAY FOURTH.

ON THE FŒTUS IN UTERO, Etc.

It is now several years since, first in a series of papers in the *Edinburgh Monthly Journal of Medical Science*, and afterwards in a separate pamphlet, dedicated to the Highland and Agricultural Society of Scotland, I directed attention to a class of facts which seem to indicate that the *fœtus in utero* may, and in fact habitually does, inoculate the female with the constitutional qualities of the male parent. I did not myself suggest the theory of inoculation. It had been advanced a short time before, in the columns of a provincial newspaper, by Mr. James M'Gillivray, a veterinary surgeon in Aberdeenshire. But I may claim the merit of introducing it to the notice of the profession, and giving currency to it among the agricultural body ; as well as of imparting to it a more scientific form, and placing it on a broader basis, than had been done by Mr. M'Gillivray.

In both the first and the second of the series of papers referred to, I pointed out the application of this principle to the question—Whether *secondary syphilis* may not thus be transmitted from the male to the female; and it is to this question that I now wish specially to call your attention, premising that my own has recently been recalled to it by two communications on the same subject, published almost simultaneously, but quite independently of each other, both supporting the same views, and both pointing to that principle of inoculation. One of these is in the *Edinburgh Medical Journal* for October last (1856), by Dr. James B. Balfour, of Edinburgh; the other is a " Report " of illustrative cases, in the *Medical Times and Gazette* for October 11th, prepared by Mr. Hutchinson, of the Metropolitan Free Hospital,* and to which is appended a summary of the principal conclusions arrived at; the report itself forming the groundwork of a paper which was read by him a little while ago to the Hunterian Society of London, and will shortly appear as an original communication in the pages of that periodical.† And my object is to bring together, as far as known to me, the facts that have been ascertained regarding it by these and other writers; and, by presenting those facts under one connected view, to aid in the further investigation of a subject which is as interesting in its physiological, as it is important in its pathological and practical relations.

---

* Now Emeritus Professor of Surgery to the London Hospital (1886).

† Since published therein—December 20, 1856, and January 10, 1857.

But (*reculer pour mieux sauter*) it will be expedient, before entering upon it, to lay before you a brief outline of the general doctrine of maternal inoculation through the fœtus, as derived from other facts, and resting on other grounds. Such an outline will, I hope, enable you the better to see clearly the bearings of our proper subject, and to appreciate its importance, not merely as a practical question, but as supplying a test or criterion of great delicacy, and of corresponding value, in regard to the whole general doctrine itself.

I. The general fact on which this theory of inoculation is founded is, that the peculiarities of a male animal that has once had fruitful intercourse with a female of the class *mammalia*, may be more or less clearly discernible in the progeny which that female may subsequently have by other males; or, in other words, that a male animal that has once had such intercourse with a mammalian female may so influence her future offspring begotten by other males, as, to a greater or less extent, to engraft upon them his own distinctive features and his own constitutional qualities; his influence thus reaching to the subsequent progeny, in whose conception he himself has had no share, and his image and superscription, so to speak, being more or less legibly inscribed upon them. Accordingly, if the female be of a *different* breed or species from that male, and have thus borne a cross or hybrid by him, her future offspring, got by males of the *same* breed and species with herself, may yet have more or less the characters of a cross or hybrid. And if it be true, as is main-

tained by Mr. Orton, of Sunderland, that, in the reproduction of the animal species, the male parent
imparts mainly to his offspring the parts and
qualities that are characteristic both of the animal
in general, and of the species to which he belongs,
as well as of himself individually—to wit, the
external structures or the organs of animal life—
the brain, nerves, organs of sense, the skin and hair,
together with the bones and muscles—and for the
most part, therefore, determines the outward
character and general appearance of the offspring—
it is precisely certain of those parts and characters
which may be expected to appear, and which are, in
fact, discernible in the subsequent offspring which
the female bears to other males.*

1. It was not till the publication in the *Philosophical Transactions* for the year 1821 (pp. 20, 23), of
the cases of Lord Morton's mare, and of Mr. Giles' sow
that the subject attracted any particular attention.
(*See Essay First ;* pp. 3, 4). These two cases were
long looked upon as solitary examples of an exceptional phenomenon, and the phenomenon itself was
long regarded as scarcely, if at all, admitting of a
plausible explanation. By some, however, it was
suggested that it might be ascribed to a partial

---

* See two papers by Mr. Orton, on "The Physiology of Breed-
ing," in the *Newcastle Chronicle* for March 10, 1854, and
November 10, 1854; and a paper by the author,—"On the
Relative Influence of the Male and Female Parents in the Repro-
duction of the Animal Species;" in the *Edinburgh Monthly
Journal* for August, 1854.

fecundation by the quagga and chestnut boar, respectively, of others of the mare's and sow's ova than those actually impregnated by them; or at least to some influence of their seminal fluid on the ova that remained, of a permanent kind, but strictly *local* in its agency, and modifying in some way the development of these ova, when afterwards impregnated by other males. By others, again, the phenomenon was regarded as an instance of that class which come within the category of "mother's marks," and are popularly ascribed to the agency of the imagination or of the mind on the body, and supposed to be illustrated by a reference to the case recorded in Genesis (chap. xxx.), of Jacob and his peeled rods. Neither the phenomenon itself, however, nor these conjectures regarding it, gave rise to much discussion, or long continued to engage the attention of physiologists. The subject was altogether passed over by Dr. Bostock, in his elaborate System of Physiology. It was referred to by Mr. Mayo and by Dr. Alison, in their "Outlines," and by Dr. Kirkes, in his "Handbook," of Physiology. But a brief and passing allusion to it was all the notice it received.

Meanwhile, instances of a like kind were becoming known to persons engaged in the breeding of horses, cattle, sheep, dogs, and swine; and in 1849 Mr. M'Gillivray adduced, in the paper I have referred to, a collection of cases sufficiently large to show that the phenomenon exemplified in Lord Morton's mare, or rather in her progeny, is by no means so unique as it was thought to be. And it may now be confidently affirmed, that since then, and within these few years, enough has transpired in this department of inquiry to warrant the presumption, that the phenomenon in question is so uniform in its occurrence, whenever the requisite conditions obtain, as to constitute a general fact or law of nature. It is so regarded by a large number of our great breeders of cattle, by dog-fanciers, and others of large experience in the rearing of horses, sheep, &c., and is habitually taken into account by them in the selection of their breeding-stock, and in their estimate of the purity of an animal's blood; the practical belief of all of them on this subject being, that a female animal that has once been impregnated by a male of a different breed or species from her own, and has borne a cross or a hybrid by him, is thereby *destroyed*, for a time at least, if not permanently, for the purposes of breeding *pure* stock of her own kind; and the *theoretical* belief of many among them, that this result happens because (as was first suggested by Mr. M'Gillivray) the blood of the female has, through the fœtus, been *contaminated* by the blood, and charged with the qualities of the male she first had intercourse with—so charged therewith (so contaminated thereby), that she herself imparts the blood and the qualities of this male to the progeny she afterwards has by other males.

2. It is not my intention to lay before you any additional examples of this kind as occurring in the lower animals. I have elsewhere brought together a tolerably large collection, and to this it must

suffice to refer. But I cannot forbear just briefly alluding to the fact, for the mention of which you will be quite prepared, and the reference to which will naturally pave the way to our proper subject— that instances of the like sort are, perhaps, equally common in our own species. It has long been known —and it is, in fact, a popular observation—that in the case of a woman twice married, and fruitful by both husbands, the children of the second marriage often resemble their mother's first husband or his family, and that not in features only, but in mind also, and in disposition. It is obvious, indeed, that in any such case, where all the individuals concerned —the woman, her children, and both the husbands —are of the *same* variety of the human family, the alleged resemblance must often be exceedingly diffi- cult to trace or to substantiate. But it is equally obvious, that means exist for ascertaining clearly whether it obtains or not. There are equally dis- tinct varieties of the human species as there are of any of the lower animals; and all that is requisite for bringing the question to a decisive issue is, to observe accurately whether the children of European parents—when the mother has, in the first instance, had offspring by a negro—exhibit traces of the latter in the colour of the skin, and still more in the quality of the hair, the form of the features, &c. ; or, contrariwise, whether the children of negro parents, when the mother has formerly been im- pregnated by a European, manifest the peculiari- ties of the latter.

Of the former case, I have heard of two instances

as occurring in this country, one of them under the notice of Professor Simpson, of Edinburgh (See *Postscript* to *Essay First,* p. 21). Of the latter case, that in which the parents are both negroes, but in which the mother has previously had fruitful connection with a European, Dr. Robert Balfour, of Surinam, wrote me some years ago to say, that repeated instances of the European influence under such circumstances, had come under his own observation. " Among the coloured population " (Dr. Balfour wrote me, April 19, 1851), " consisting of negroes and their offspring by Europeans, the most striking illustrations of the truth of your, or Mr. M'Gillivray's, theory are constantly occurring. Shortly after my settlement here, I was led to remark the circumstance (which I had never noticed recorded by writers), that if a negress had a child or children by a white, and afterwards fruitful intercourse with a negro, the latter offspring had generally a lighter colour than the parents. . . . I could particularize many such, the observation of which had led me to ·form an opinion similar to that which you have more elaborately worked out." Doubtless, if looked for, examples of the influence now in question would be found as general in our own species as in the lower animals. And it may just be observed that, in the one as in the other, they can only be unequivocally and satisfactorily determined in cases where the individuals or the animals that are the subjects of observation are of a different race or species; or rather, where the male, the female, and the offspring, are of the same breed and species, but the female has first of all been impregnated by a male of a different breed or species.

3. In ascribing this very remarkable phenomenon, as seen both in our own species and in the lower animals of the class Mammalia, to an agency exerted by the fœtus on the mother's system, of the nature of inoculation, I do not intend to consider in detail, or in connection with other explanations that have been given of it, the grounds on which that doctrine of inoculation rests. The questions involved in it are too numerous, and the discussion of them would be too tedious, to admit of my entering upon them at present. I have elsewhere considered them pretty fully, and must here content myself with this general remark, that the facts there advanced in support of the theory of inoculation have always seemed to me sufficient, if not indeed absolutely to prove the truth of it, to impart to it at least a probability short only of absolute certainty : and, further, that while I deemed it expedient to suggest to the agricultural body certain experiments in breeding, which, if carefully made, could not fail to be decisive of it, I have always thought that a sufficiently large collection of facts, of the description of those now to be laid before you in regard to the poison of syphilis, would supply all that is really wanting to make the *demonstration* of its truth complete. The reasonableness of the doctrine no one can well dispute. Connected as the fœtus and its mother are by the placenta, a mutual interchange of fluid matters is continually going on between

them : the fœtus drawing supplies from its mother's blood for its growth and maintenance, and the mother (though, no doubt, in much less quantity) abstracting materials from the blood of the fœtus. Is it unreasonable to suppose that the matters abstracted by the mother may be charged with (or have inherent in them) the *constitutional* qualities of the fœtus; or that passing into her body, and mingling there with the general mass of her blood, they may impart those qualities to her system? And this supposition will perhaps appear the less improbable, if regard be had to the length of time during which the connection between the mother and the fœtus is kept up, and during which this transference of materials must go on—a period of some weeks, or even of several months. But the qualities inherent in the fœtus must in part be derived by it from its male parent, and be to that extent identical with his. The distinctive peculiarities, therefore, of this parent may thus come to be *engrafted* on the mother, or to attach in some way to her system; and if so, what more likely than that they should be communicated by her to any offspring she may afterwards have by other males ?

4. Hitherto we have had under consideration the transmission of such qualities only as are strictly normal or healthy, and so far the subject may be regarded as one of physiological interest merely, or at least of practical importance to breeders alone. But it is conceivable that *morbid* qualities may be transmitted also, and be in the same way, or

through the same channel, engrafted on the mother's system. In our own species, scrofula may, peradventure, be thus transmitted. In my First Essay on this subject, I not only refer to the possibility of this, but give two instances in which it apparently happened (pp. 10, 21). A serious difficulty, indeed, attaches to the investigation of this point. So *widely* diffused is the scrofulous diathesis among mankind, that in any single case in which it has apparently been acquired in this way, it would be difficult to determine whether it was *really* thus derived or not. Still, observations on the large scale, long and carefully conducted, as to the health and constitutional tendencies of females before and after marriage, as compared with those of their husbands and children, and carefully distinguishing between females that have and females that have not had children, might in the end, and after full allowance for errors or fallacies of observation, sufficiently demonstrate the reality of the occurrence. Certainly it is not an uncommon observation, that women born of healthy parents, and themselves previously healthy, fall into a state of general ill-health after becoming mothers*—an observation which may yet be connected, in a certain proportion of cases, with the principle now in question. So, likewise, may the syphilitic poison, in its *constitutional* forms, be transmitted

---

* Even this observation has not escaped Mr. Kingsley. Speaking of Mrs. Vavasour, he says—"And when children, and the *weak health* (on the part of the mother) which comes with them," &c.—*Two Years Ago*, vol. i. p. 70.

after the same manner from the husband to the wife, and be by her conveyed to the children she may subsequently bear to a second husband, himself perfectly free of all syphilitic taint. Whether any *other* morbid states besides these—*e.g.*, mania, epilepsy, gout, &c., which are unquestionably hereditary—may thus be imparted and transmitted, it is at present impossible to say. Enough appears, however, from a consideration of the transmission of healthy qualities, and from what is known in regard to the transmission of syphilis, to impart to the whole inquiry as to the transmission of morbid qualities, an interest of the highest practical kind, and that in a *social* as well as in a purely *professional* point of view. So important may it yet be ascertained to be, in contemplating marriage with a widow, to have regard to the constitutional peculiarities of the deceased husband.

So much for the general outline I spoke of at the outset—fuller, indeed, than was intended, but, perhaps, not more so than it was necessary to premise in order to the right understanding of our proper subject—the fœtus in utero, as the channel of transmission of secondary or constitutional syphilis from the male to the female parent.

II. Taking up, now, the subject of *syphilis* in connection with the alleged inoculating power of the fœtus, let me just observe at the outset, that there are two points of view from which it may be regarded—two objects with which its investigation may be conducted :—*First*, as a part of the history of syphilis, and in relation to pathological and practical ends; and, *secondly,* as a test or criterion of the general doctrine of inoculation itself, its special value as a test lying in this, that nothing like *mental* influence can possibly be imagined to have any share in the transmission of the virus. And these two objects, although so far distinct, need not—nay cannot—be separated, in the prosecution of our scientific inquiries. That is to say, if such inquiry shall demonstrate that the syphilitic poison in its constitutional form may be, and actually is, transmitted—nay, is only communicated, and is communicable only, from the husband to the wife through the fœtus—while this fact would form a real, a practically important, addition to our knowledge of the laws of that poison, it would, at the same time, establish the general doctrine of inoculation, as exemplified in other modes, and resting on other grounds; and it would impart, besides, a peculiar strength to the evidence in support of it thus derived, because absolutely devoid of all leaven of the fallacy which may be supposed to attach to that and every other source of evidence.

1. The *first* question I shall consider, as more obviously connecting itself with the general features of the cases already referred to, and exemplified in Lord Morton's mare, is—Whether a woman, twice married—first to a constitutionally syphilitic husband, next to a husband untainted by that poison— and fruitful by both, may herself (not otherwise tainted than through her former husband) infect the children of her second marriage ?

In seeking formerly (in 1849) for a solution of this question, which I thought I had been the first to suggest, I was unable to do more than adduce in illustration of it an imperfect observation from Messrs. Maunsell and Evanson, to the effect, " that they have notes of the case of a syphilitic child, whose mother had been infected by a former husband, and to all appearance cured five years before its birth ; the father of the child, her second husband, being in perfect health."* They do not say, however, in what way she was infected ; and, as it may have been in that of primary affection or chancre, the case, as it stands, is of no value in relation to the question before us.

Dr. Montgomery, of Dublin, I afterwards found, had anticipated me in this question as to syphilis. Referring to Lord Morton's mare, and Mr. Giles' sow, Dr. Montgomery remarked, in 1837, in the first edition of his great work " On the Signs and Symptoms of Pregnancy :"

" Such occurrences appear forcibly to suggest a question, the correct solution of which would be of immense importance in the history and treatment of disease. Is it possible (he asks) that a morbid taint, such as that of syphilis, for instance, having been once communicated to the system of the female " [by a conception ?] " may long linger there, and influencing several ova, continue to manifest itself in the offspring of subsequent conceptions, where impregnation has been effected by a perfectly healthy man, and the system of the mother appearing to be at the time, and for a considerable period previously, quite free from the disease ? My belief" (he adds) "is certainly in favour of the affirmative."

---

* On the Management and Diseases of Children, 5th edit., pp. 452, 453.

What Dr. Montgomery here suggested as a probable occurrence, and then merely suspected, he has since been able to substantiate as a fact :—

"Such " (he observes in the second edition of that work, published in 1856) "was the opinion I expressed in 1837 ; and further experience and observation have, I think, shown to be a fact what I could then only venture to say I believed to be likely."

He then proceeds to quote two cases in support of this observation—one from M. Vidal, the other from M. Cazenave—remarking afterwards that, while he thought the observation of the fact was original with himself, he had found it was made long ago by others—by M. Gardien, in 1824, and by Dr. F. H. Ramsbotham, in 1835.* In the case given by Cazenave,† the syphilitic taint was trans-

---

* We shall see by and by how clearly Dr. Ramsbotham had divined it. It seems to me, however, very doubtful whether M. Gardien had any conception of it whatever, judging at least from the passage referred to by Dr. Montgomery, and which is as follows, "Le moment de la conception devient quelquefois une circonstance qui favorise chez la femme le developpement de differens virus qui étaient restés sans action jusqu'à ce moment. L'état de grossesse manifeste quelquefois des symptomes de virus vénérien qu'on n'avait pas encore aperçu, et qui, sans cela, aurait pu rester encore plus ou moins de temps engourdi. Le virus syphilitique, avant la conception, pouvait circuler dans la masse generale sans affecter aucune glande, et sans donner de marques de sa présence : parceque ces organes n'étaient pas sensibles à son action. La grossesse augmentant la sensibilité, manifeste des symptomes d'une infection générale, parceque les organes glandu leux ressentent, à cette époque, les impressions du virus vénérien, auquel ils étaient insensibles auparavant." ("Traité d' Accouch." 1824, vol. ii. p. 29.)

† Traité des Syphilides, etc., p. 133.

mitted by the mother to the children of her *third* husband, as well as to those of her second; but the case is essentially defective, inasmuch as it is not stated in what manner the woman *originally* became affected. It would even appear from the narrative of the case, that it was while a widow that she became affected: and as it does not appear that she had a child subsequently till she married her second husband, it is probable that she acquired the disease in the primary form, at least otherwise than through impregnation. M. Vidal's case, however, is in every respect unexceptionable, and exactly to the point:—

" A woman, whose husband was affected with constitutional syphilis, gave birth to a child, which in two months showed symptoms of that disease, of which it died. The woman never had any appearance of syphilitic affection, not even sufficient to soil her linen. Her husband died, and, after remaining some time a widow, she married a healthy man; and about twenty months afterwards, being four years after the former birth, she bore a child, which in two months presented the same form of syphilitic eruption which had appeared on the former child."

It is plain that opportunities for observing cases of this description must be comparatively very rare. A second marriage, indeed, on the part of widows, is not uncommon; but with the conditions requisite for thus tracing the transmission of the syphilitic poison, very uncommon. And it must be kept in mind, that in some cases of *this* kind, the results of observation (although, as we shall see, of great value notwithstanding) will be of a *negative* character, as in the following case given by Dr. James B. Balfour:—

" A respectable young woman from the north was married to a tradesman, who had no trace of syphilitic disease at the period of marriage ; but he afterwards acknowledged that, two years before, he had the disease, followed by slight secondary symptoms, which had entirely disappeared under medical treatment, and he had seen or felt nothing since. This woman complained of nothing until about three months after she became pregnant. Then, however, symptoms of secondary syphilis became apparent—spots of psoriasis appeared on various parts of the body—hard knots were felt on the perineum, and on the external labiæ—and within the vagina there was a hard, knotty feeling over the whole mucous surface. Her child exhibited a distinct syphilitic appearance, which was removed by the usual treatment. Shortly after delivery all symptoms of syphilis entirely disappeared ; and as she shortly afterwards removed to the country, she was subjected to no medical treatment. A few months afterwards her husband died. She subsequently married a farmer. About six or eight months since I saw her, and she has borne three children to her second husband, and certainly more healthy children I could not wish to see. She informed me that she had never been under medical treatment since I had attended her, indeed she never had a medical man near her, except during her confinements, and that she had never suffered, during any of her pregnancies, from anything like what she had done at the first one."

I shall not trouble you again with the details of individual cases. But I have thought it desirable to quote at length the two I have taken from Dr. Balfour and M. Vidal, as being fair examples of the kind of cases we may expect to meet with, and on which our conclusions must be founded ; few of them exactly tallying in every point ; most of them having each its own special features, and the comparative value of the several cases being widely different—none, however, being of any value in which it does not clearly appear that the virus was imparted to the female in its secondary or constitutional form only. In both the cases adduced this essential condition was plainly apparent, but they differed in other respects. In M. Vidal's the woman never exhibited in her own person any symptoms of syphilis, but gave proof of being tainted with it by giving birth to a syphilitic child, got by her second husband, a healthy man. In Dr. Balfour's this evidence was wanting; but the fact of the woman having imbibed the poison from her first husband, appears from the symptoms of it having shown themselves upon herself during her first pregnancy : the evidence thus afforded deriving confirmation from this other fact, that the child she then bore was also affected with it.

2. And this, for the most part, is the *sort* of proof we should look for—direct evidence furnished by positive manifestations in the person of the mother, and confirmed by like manifestations in that of the child, of syphilitic taint. Although always interesting, and indeed important to do so, it will not actually be necessary, except in a very few cases (as in M. Vidal's), to seek to trace the transmission of the virus to the children of a second marriage, and of a healthy father. The proof we require lies within a narrower compass, and may be reached by a shorter process. Dr. Balfour's case rather than M. Vidal's, is the type of the cases we have to search for and to consider ; and, fortunately for the investigation, they are of much more frequent occurrence. And the question to be considered, in the first instance, is simply this—Whether ever a woman derives secondary syphilis from her husband, *when* she conceives by him, and *because* of her doing so.

The evidence now available on this point, as far as known to me, is of two kinds :—*First*, General statements of experience supplied by M. Ricord, Dr. Montgomery, Dr. Tyler Smith, Dr. Carpenter, and likewise by the author cf the review, in the *Edinburgh Medical Journal*, of the new edition of Dr. Montgomery's work, and by Dr. Ramsbotham : and, *secondly*, Specific instances observed and recorded by various writers.

Let us first consider the former branch of the evidence. M. Ricord, one of the highest authorities on all points relating to syphilis, observes that his experience goes to prove, that in the case of a woman pregnant of a child whose blood is contaminated with syphilis acquired from the father, this child may, and often actually does, contaminate the mother's system :—

" So long as a diseased father (says M. Ricord) is under the influence of constitutional syphilis, the germ which is by him conveyed into the uterus carries along with it the syphilitic diathesis. . . There is no such thing as an infection of the child by the mother, she having been contaminated by the father ; but the husband procreates an infected child, which may then propagate the secondary poison to the mother—*for where there are no children the mother does not suffer.*"*

Dr. Montgomery, in the last edition of his " Signs and Symptoms of Pregnancy," speaking of this very point, remarks, that " there can be no doubt of the *frequent* occurrence of the fact." Dr. Tyler Smith's testimony is, that the most common mode in which

* The Lancet, April 8, 1848, p. 383. See also Mr. Acton's work on Syphilis, p. 632.

women become affected with syphilitic *uterine* disorder, is, he believes, that in which the fœtus is the medium of communication.* Dr. Carpenter, speaking of the inoculation theory, says :—

" This idea is borne out by a great number of important facts ; and it serves to explain the circumstance, *well known* to practitioners, that secondary syphilis will *often* appear in a female during gestation or after parturition, who has never had primary symptoms, whilst the father of the child shows no recent syphilitic disorder."†

And the reviewer referred to observes—

" That a woman may get syphilis from a man without any external disease—namely, by bearing a syphilitic child in the womb of her previously healthy body, which child infects her with the disease : and that she may then exhibit outward signs of the disease now in her constitution—or, without this, may prove her syphilitic taint, by producing syphilitic children to another perfectly healthy father."

Let us see, further, on this branch of the evidence, what Dr. Ramsbotham says. Among the first, if not the first to suggest the possibility—nay, the great *probability*—of this mode of transmission of the syphilitic poison, nothing can be clearer or more explicit than his observations on it :—

" It is a generally received opinion, I believe, that syphilis in its secondary stage, is not communicable directly to either sex from the other—that the disease is not propagated unless there exist an open chancre ; and this accords with my observation. But it appears to me *probable*, that if a previously healthy woman conceive of an ovum tainted by syphilitic virus derived from its father, her system may become inoculated during the progress of

---

* See Association Medical Journal, July 14, 1854.
† Principles of Human Physiology, 5th edition, p. 826.

gestation, in consequence of the close vascular connection existing between it and herself ; for it has fallen to my lot *to see more than one case* in which a young woman, united to a man labouring under obstinate secondary symptoms, remained *healthy* for *some months after marriage,* but became the subject of the same disease in its secondary form *soon after impregnation had taken place ;* and *I have considered* that, in such a case, the mother derived the disease, *not directly from the father,* but *from the affected infant* which she carried in her womb."[*]

These are strong and unequivocal statements, and the specific evidence furnished by individual cases is equally decided. To what extent the records of medicine can now supply examples of this description, I cannot say. But I have already repeatedly alluded to those brought forward by Dr. Balfour and by Mr. Hutchinson ; and it is to these that I would now particularly refer. The former gives four or five cases, the latter as many as fifty. To be duly appreciated, these cases must be examined individually and compared together—a task which it were foreign to my object to enter upon at present, even were it possible to accomplish it without actually reproducing the cases in detail. It must suffice, therefore, thus to refer to them, and to remark that, having carefully considered them, I think they conclusively establish in the affirmative the question presently before us ; proving that a woman may, and often actually does derive syphilis from her husband, when she conceives by him, and in consequence of her doing so.

3. The next question to be considered is—Whether at each pregnancy there may be a *renewal* of the

---

[*] Medical Gazette for May 23, 1835.

symptoms of secondary syphilis in the woman—implying that she has from some source, or from some cause connected with the state of utero-gestation, imbibed a fresh dose of poison ?

On this point, and speaking apparently of uterine syphilitic affection, Dr. Tyler Smith stated at a meeting of the Medico-Chirurgical Society, that " he had observed in such cases, that at each pregnancy a fresh dose of the syphilitic poison is imparted to the mother, unless in the meantime the husband had been the subject of anti-syphilitic treatment."* Mr. Hutchinson's experience as regards the ordinary secondary symptoms is to the same effect. Our time will not allow me to refer specially to his cases, as bearing on this question ; but it will suffice to quote his own general conclusion :—" Increase of symptoms and relapses may be produced by repetition of exposure to contagion "— *i.e.*, by the woman again becoming pregnant. And in one of Dr. Balfour's cases, the first in his collection, there was a recurrence of the syphilitic symptoms during the second pregnancy.

The positive value of the fact thus clearly ascertained, and still more, its value in relation to other facts, and as checking or testing these, I am disposed to rate very high.† The consideration of it, how-

---

* See Association Medical Journal for July 14, 1854.

† As this point is important, I hope I may be excused, if, in order to bring it fully into view, I adduce here so much of Dr. Balfour's first case—the only one in which relapse occurred—as is illustrative of it :—" By the time the lady recovered from her confinement, and the lochial discharge had ceased, all trace of the

ever, in this view, must be reserved till another question yet to be proposed comes before us. Nevertheless, to meet an exception that may be taken to it, it will be necessary at this stage to point out how the fact really stands. The reappearance of the syphilitic symptoms in the mother during a second or subsequent pregnancy, seems certainly to imply that she has received a fresh dose of the virus, and likewise that she has received it directly from and through the fœtus. But it does not absolutely follow that in every such instance the fœtus derives the virus from the father. In most instances it no doubt does. M. Vidal's case, however, is sufficient to show that it may derive it from the mother exclusively; and in as far as it does, the reinfection of the mother comes remotely from herself. But this does not alter the character of the fact, or impair its value. Why, while long previously dormant in her system, the virus should thus again become active and reappear, we shall

---

syphilitic affection had entirely left, and she was so completely restored to health that I did not deem any medical interference necessary. She enjoyed excellent health, and made a good nurse, and continued quite free from any return of the disease *for more than fifteen months,* when she again became pregnant. About two months after this occurred, a train of symptoms precisely similar to the first appeared, and continued as formerly up to the time of her confinement, when they again disappeared as on the former occasion. The second child was also syphilitic." It cannot well be supposed that during the interval between the two pregnancies, no sexual intercourse obtained between this lady and her husband till just previously to the second; and that the relapse and the reimpregnation were merely coincident.

afterwards consider; merely remarking at present that the like question may be raised, and admits of a like solution, in regard to the power of the virus when long dormant in the father's system to infect the child.

4. Another question still remains—Whether ever a woman derives secondary syphilis from her husband unless she conceives by him ? And the question to which this stands opposed is—Whether she may not derive the virus through the *seminal fluid* also, and indifferently *at any time ?*

This question, which was suggested to me by Mr. Paget (Sir James), as it was, I believe, to him by the perusal of my own observations on this subject in my first paper, is perhaps the most important of all.* Previously, however, to taking it up, the consideration already referred to, as to the power of a seemingly *dormant* virus, demands attention. And it does so as enabling us to understand why— if the fact be so—it is through the fœtus only, and

---

* The alternative question might have been framed in more general terms so as to include tainted *preputial* as well as *urethral mucous secretions*, together with exudations from vesicular or pustular eruptions on the penis. But I have restricted it to the seminal fluid, in accordance with Mr. Paget's communication. "I would venture to suggest," Mr. Paget wrote me, "that you should try to find whether ever a woman derives secondary syphilis from her husband *unless she conceives by him.* Facts bearing on this point might prove that secondary syphilis is not communicated directly by the *seminal fluid,* but by the child begotten with it ; and this mode of inoculation being proved, would go far to prove the foundation of your theory "—*i.e.* Mr. M'Gillivray's.

during pregnancy alone, that the female derives the virus from her husband, or, in a few cases, virtually reinfects herself; and, likewise, in the absence of positive proof, and until such proof is adduced, that she may and does, making it *probable* that she will not and cannot derive it through the seminal fluid, and indifferently at any time.

Now, although in many cases, in the lower ranks of life particularly, the syphilitic symptoms in the husband may continue to show themselves, and be severe or well-marked even after marriage, the general fact, I apprehend, is, that among all classes, and certainly in the higher, they have long previously disappeared; that in a large proportion of the cases they were originally mild, or very slight in the later periods of their manifestation; and, further, that, to all appearance, the virus itself has been completely eradicated from the system. Facts, indeed, clearly demonstrate that it must still be present there. But facts demonstrate, also, that it must be present in such quantity and in such condition as to be incapable of again exciting actual disease in *him*. And the question arises—how, under such circumstances, should the virus be capable of tainting and sensibly affecting the *child* begotten by him? The answer is not far to seek. In accordance with the known laws both of physiology and of the morbific poisons, the fœtus, being an entirely new formation, and the seat and subject of nutritive changes of great activity, will be *peculiarly susceptible* of the virus; and the virus itself, although existing in the father's system, and imparted by him

to the fœtus in a quantity which may be infinitesimally small, acting as a *ferment*, multiplies itself in
the blood of the fœtus, and at length *accumulates*
there in such quantity, acquiring at the same time
such *efficiency* as to produce in it *manifest* syphilitic
disease.   The like explanation will apply to the
cases in which (as in M. Vidal's), the virus is derived
by the fœtus from the mother.   And that the fœtus
thus contaminated, whether by the mother or the
father, and thus actually affected by the poison,
should readily infect or reinfect the mother, is only
what might reasonably be expected.   Thus, strictly
speaking, it is not the father that *directly* affects the
mother.   It is the fœtus that does so.   The virus,
in passing from the father to the mother, passes
first into the system of the fœtus, and there multi-
plying, acquires the requisite power.   And it may
be, and I confess I cannot but believe that it is
thus only that, while *latent* in the system of the
husband, or appearing only in the *slight* and *mild*
tertiary form, it ever *will* or *can* affect his wife.

To revert now to the questions put a little ago.
We have already seen that the female may, and
often does derive syphilis from her husband when
she conceives by him, and that—whether remotely
derived now from herself or not—she may, and
often does *again* exhibit the like symptoms on a
subsequent impregnation.   But may she not also do
so independently of conception, and at any time ?
May not sexual intercourse alone, and the mere
application of the seminal fluid, or of tainted preputial secretions, suffice to impart the virus to her ?

I am inclined to think not. It were a rash assertion to say that its occurrence is impossible, and one at least which it would require a large amount of negative evidence to make good. We know that a syphilitic infant may infect its nurse, and it is clear that the infection must come to her through her own nipple and the infant's mouth—abrasion or fissure of the nipple being probably a condition essential to the result. In this case, however, we have an infant labouring under *actual* syphilitic disease; and if infection be possible in the case of man and wife without impregnation, I apprehend it is so only in cases in which the man is at the time *actually diseased*, and of these, probably in those only, in which the female is labouring under ulceration or abrasion of the os uteri. But that it is possible in cases in which the syphilitic affection in the man is altogether latent, or shows itself in its slighter forms, and the mucous surfaces in the woman are entire—I do not believe. Yet it is in cases of this description that it is so often seen in connection with pregnancy—the woman healthy, and the man without any ostensible syphilitic affection.

The question, however, is one of fact. A single well authenticated and unequivocal instance of infection occurring independently of conception, would establish the infecting power of the seminal fluid; or else of certain others of the male genital secretions. Have any such cases ever been met with and recorded? Dr. Balfour refers to none, and seems to consider it certain " that the mere fact of coitus between the woman and her husband is not sufficient to produce the disease." Mr. Hutchinson states that he has met with but one case of this class; but he regards it as a doubtful one, and he holds it to be "extremely doubtful if 'contagion by the seminal fluid' be possible"—observing "that cases are extremely rare in which married women who have never conceived, become the subjects of constitutional taint without having had primary symptoms, and are in all probability to be explained as errors of observation." And, as we have seen, M. Ricord states emphatically, that "when there are no children the mother does not suffer." I have not myself met with any, although for nearly eight years I have had my attention directed to cases of this description ; nor have I been able in the course of my reading to find any described. It is true that writing in 1850, I said—"I fear it will turn out on inquiry, that secondary syphilis may be transmitted directly by the seminal fluid, independently of conception." But this apprehension was founded, contrary to my hope, on answers I received to inquiries made of some professional friends of experience in this department of practice, who *thought* they had seen cases in which the female had been tainted by her husband, and (the idea of inoculation through the fœtus being new to them, and appearing strange) were persuaded that the contamination *must needs* have occurred through the seminal fluid. To the apprehension, however, then expressed, I appended this remark :—"but perhaps it may appear also that its transmission in this way (*i.e.*, by the seminal fluid) is *occasional* only and *uncertain*, while it is *very frequent*, or almost *inevitable*, when conception follows intercourse "—adding, that "a comparative observation of this kind, if clear and undoubted, would be nearly equally decisive." —*Essay Second*, p. 43.

Now, however, without raising any question as to the abstract possibility or impossibility of maternal inoculation through the semen of the male parent, or otherwise by him independently of impregnation, we are, I submit, in a position to say confidently,— and that mainly through the labours of Mr. Hutchinson, Dr. Balfour, Dr. Ramsbotham, M. Ricord, and others, *first*, that no evidence of a reliable kind has yet been adduced to show that inoculation apart from conception, or otherwise than concurrently with it, has ever occurred, or been known to occur; and, *secondly*, that inoculation by the fœtus—never suspected until within these last few years, is not only possible, but is at once the special mode in which it is effected, and the only mode in which it is known to be effected,—while it is of very frequent occurrence besides.

Nothing, I think, can now set aside these conclusions. Further observations may modify them; but the evidence on which they rest is large enough and clear enough to enable us confidently to rely upon them. And its strength lies not merely in the number of instances in which different females have become infected with secondary syphilis when they conceived, *and then only*, but in which the same female has been reinfected simultaneously with re-impregnation, and *at no other time*—continuing *free from* syphilitic affection *for months together*, although cohabiting with the tainted husband, and then *suffering again on again* becoming pregnant. It is this which makes the evidence not merely convincing but irresistibly conclusive. Conclusive it is in a twofold sense—a negative and a positive sense. Intercourse between husband and wife to any extent, or for any length of time, *harmless*, as long as there is no specific result from it. But no longer. Impregnation effected, inoculation also effected. Cumulative evidence of this kind may, in truth, be said to have already effectually set aside all notion of inoculation by the male semen, or in any way save and except by impregnation.

III. This properly concludes what I have to say on the *fœtus in utero* as the channel of transmission of secondary or constitutional syphilis from the male to the female parent. Practically important on its own account as the subject is, it is physiologically important also, as a test or criterion of the general doctrine of inoculation through the fœtus. Did our time permit, I would endeavour to point out its bearing in this relation, and its peculiar value. I would proceed to show you that there probably are in nature two sets of cases, or two classes of facts, connected with the peculiarities seen in the offspring of mammalia, *analogous*, nay, *identical* in their *external* characters, but widely different in their origin and in the conditions of their production ;

the one referable to *mental* states in one or other of
the parents, but oftenest to mental states in the
female, and operating on her either at the time of
conception or during the early stage of pregnancy ;*
the other to a change effected in the constitutional
and reproductive powers of the female by a *physical*
agency, originally extrinsic to her, but inherent in
a former fœtus by derivation from its male progenitor, and conveyed to her in the way of inoculation by that fœtus while *in utero*.

To establish this, and to demonstrate also the
relative proportion in which the two sets of cases
obtain, together with the comparative efficacy of
their respective causes, and the conditions that are
essential to the action of each, would be an achievement in physiology of great general interest ; and,
in relation to the practice of medicine on the one
hand, and the breeding of stock on the other, of
real practical importance.

But, in the investigation of one of the branches
of this subject, the first and essential point is to
make good the law or principle on which it rests—
to be well assured that the foundation is secure.
And it is in this point of view that the inquiry as
to syphilis is so important. As regards *it*, there
can be no room for suspecting that *mental* causes
can possibly be concerned in the transmission of
the virus from the father to the mother. A child
born into the world with unmistakeable manifesta-

---

* See *ex. gr.*, the case of the mare and horse (a gelding)
narrated in *Essay Third*, pp. 60, 61.

tions of syphilis on its body, begotten by a healthy man, but the fruit of a second marriage, and the first husband and the children of this marriage syphilitic also, could not for a moment be supposed to have come by it *per fortem imaginationem* of its mother, as it might be, if, instead of syphilis, it had the nose or the features of its mother's former husband. As little likely is the mother, by such an effort of mind at the moment of conception, thus to contaminate herself.*

Our time, however, will not allow me to do more than to indicate in this brief manner the value of the syphilitic test of the inoculation *per fœtum* theory; and to add, that, while it seems to me to establish it beyond the reach of all doubt or cavil— the great extent to which recent observations have shown the general fact exemplified in Lord Morton's mare to hold among cattle, sheep, dogs, horses, and others of our domestic animals, as compared with the limited extent to which mental causes are seen to influence the development of offspring, sufficiently

---

* There seems to be with some as little limit to the physio-logical agency of the imagination as to their own exercise of this faculty. In a case of legitimacy that came before it, the parliament of Grenoble went as far as to declare the issue legitimate, although the husband had been four years absent from his wife, on the ground of an allegation by the latter, that she had conceived through the power of the imagination alone :—" Ut mulier per fortem imaginationem putaverit se in somnis rem habuisse cum marito, atque sic concepisse." Had the child, however, been born syphilitic, it is possible that the parliament might have come to a different decision—straining at a gnat, albeit making no diffi-culty of swallowing a camel.

demonstrates that in *the great majority* of cases of this description, the *inoculation* principle is that concerned in their production.

IV. There is yet another test of this inoculation principle, for the suggestion of which, as of the former, I am indebted to the sagacity of Mr. Paget. The phenomenon exemplified in Lord Morton's mare *ought*, by the theory, to be *restricted* to animals of the class mammalia, that is to animals in which the young are developed within the body of the mother, and between which and the mother there exists a placental connection during the whole period of intra-uterine life. But is it actually so restricted ? May it not be seen in *birds*, in which the young are wholly developed outside the mother's body, and between which and the mother there is no placental connection ?

Now, among bird fanciers crossing is perhaps as extensively practised as it is with our domestic quadrupeds. Nothing, for example, is more common than to breed the barn-door hen with the bantam cock one year, and the next year with a cock of her own kind ; or the hen-canary one year with either the cock-linnet or the cock-goldfinch, and the next with the cock-canary. In these and other cases of the like kind, cross birds or mules are obtained the first year ; and these, in accordance with the law laid down by Mr. Orton, of Sunderland, take very strongly after the cock. But are the birds obtained the following year, when the parent birds are both of the same species and of the same breed, perfectly *pure*, that is, without taint or trace of any of the characteristics of the male bird with which the hen was bred from the previous year ?

On this point I made very extensive inquiry several years ago in quarters where information was to be had, carefully concealing the object I had in view in making the inquiry ; and the uniform result was, as I stated in 1851, in my pamphlet " On a Remarkable Effect of Cross-Breeding,"—that no trace or admixture of those antecedent male birds is ever seen in subsequent broods, the birds produced being in every instance as pure as in those in which no such crossing had been previously effected with the mother-bird.

It is but right, however, to state that two different individuals, writing quite independently of each other, have alleged that the reverse is the true state of the case. One of these is Mr. Orton, of Sunderland ; the other, the reviewer in the *Aberdeen Journal* newspaper (February, 1851) of my pamphlet on Cross-Breeding.

The latter, quoting my remarks on this point in the pamphlet, says :—

"He gives Mr. M'Gillivray's theory one advantage which is not its due. So far from its being the case that the peculiar effect of crossing here in question, fails to be seen in birds, it is in them that it is most of all apparent. It is well known that a hen-canary is rendered useless for breeding pure canaries by crossing for mules. After a cross with a goldfinch or linnet, no peculiar feature of these birds may turn up in subsequent broods by a male bird of the hen's own species, but every nest will be discoloured ; and so long as the hen produces young, it will be remarkable indeed, if a clean yellow canary be found among her progeny. This," he adds, " may easily be certified ; and we are inclined to think that it must be regarded, as in a great measure

destructive of the theory of the purely physical cause of inoculation."

And so it would *if true*, but for the syphilitic
test. This, it seems to me, places the theory in
question beyond all doubt. But without speculating
at present on the conclusions to be drawn from the
fact, if the case be as this writer represents it, let us
first consider whether it be a true or a " false " fact.
Now, on reading the statement I have quoted, I
sought and obtained a personal interview with the
writer, and found that he could speak much less
confidently than he had written, and, in particular,
that he wrote from hearsay, and without any personal knowledge or experience in the matter. Having
ascertained this, I subsequently renewed the inquiries
I had formerly made, with the same precaution as
before to conceal the object I had in view, but among
a larger number of bird-fanciers, and with respect
to a much greater variety of the bird tribe. The
result of this inquiry was, in every instance, the
same as that of the former inquiry. None of my
informants had ever seen anything that corresponded
with the statement of the writer in the *Aberdeen
Journal.*

More recently Mr. Orton, taking exception to the
inoculation theory altogther, observes :—

" The hen does not carry her offspring *in utero.* Notwithstanding, her offspring, *as we have already seen*, are as liable to be in-
fluenced by the action under discussion (*i.e.*, by crossing) as are
those of mammalia."

And again—

" Hens and their offspring have not this uterine connection ;
consequently, they cannot undergo the process of inoculation laid

I

down, yet they are, equally with mammalia, the subjects of this peculiar law of breeding." *

In the latter statement, Mr. Orton speaks absolutely; in the former, from an instance in point previously referred to, or adduced in some part of his paper. In all probability the *absolute* statement is founded on the *specific*, and to be judged of by it. Now, on referring to the instance which proves the ground of the latter, it is seen to be such an instance as would not be relied on, or regarded of any value whatever, in the case of a mammal. It is a case, not in which a hen-bird was bred from with a cock of her own breed, after having previously been bred from with a cock of a different species, but the reverse of this—a bantam hen having been bred from with a cock silk-fowl, and producing cross birds, after having been bred from the previous year with a bantam cock; and the observation relied on by Mr. Orton as fatal to the inoculation theory was, that some of the birds of this brood bore a greater resemblance to the bantam, and less to the silk-fowl than did others. But this by no means warrants the inference, that the more decided resemblance of them to the bantam was due to the influence of the bantam cock of the preceding year. And I will venture confidently to affirm, to take a case in which a considerable number of young are thrown off at each litter, that, if a sow of a particular breed and colour be

---

* On the Physiology of Breeding, in the *Newcastle Chronicle*, Nov. 10, 1854.

served with a boar of a widely different colour and breed, the sow never having been bred from before, the litter will be *mixed;* some of the young pigs taking chiefly after the boar, others of them chiefly after the sow. This test, therefore, which may be called the *bird test,* is thus seen to be of no value whatever. It tells nothing against the inoculation theory. Nay, it serves, negatively, to add confirmation to it, and to lend support to the syphilis test.

I have thus endeavoured, as impartially as one partial to the inoculation theory well could, to lay before you a complete general view of this interesting and important subject. I have referred to every fact known to me as bearing upon it, and have omitted nothing that seems in any way adverse to the theory in question. And submitting, in particular, the *bird test* and the *syphilis test* to the consideration of those among you who may take an interest in that theory, I will conclude by quoting a suggestion made by Dr. Carpenter in regard to the latter, but in its principle as applicable to the one test as to the other :—" As this is a point of great practical importance, it may be hoped that those who have the opportunity of bringing observation to bear upon it, will not omit to do so."*

---

* Principles of Human Physiology, 5th Ed., p. 826.

APPENDIX.

—o—

I. In the appendix to Essay First, I referred to a remarkable statement made by the excellent Count de Strzelecki, as to the effect of fruitful intercourse between the aboriginal female of certain countries and the European male, in rendering the female ever after sterile with a male of her own race. The Count's statement, founded on " hundreds of instances " coming under his own observation, unqualified by the observation of a single exceptional instance, and bearing on the aborigines of Canada, of the United States, of California, Mexico, the South American Republics, the Marquesas, the Sandwich and Society Islands, and those of New Zealand and Australia, is, that " whenever such intercourse takes place, the native female is found to lose the power of conception on a renewal of intercourse with the males of her own race, retaining only that of procreating with the white men."

Forbearing to trouble you with the application I made formerly of this statement, on the assumption of its being the expression of a general law of nature, to the theory under consideration—I think it right to mention with regard to the statement itself, that, while I then suggested the propriety of " keeping the mean between the two extremes of too much stiffness in refusing, and of too much easiness in admitting" it to be a matter of *fact*, I have since received such information as warrants me in refusing to admit it as being *universally* true, at least in respect of certain of the races specified by the Count de Strzelecki. My friend and former pupil, Dr. William Sim Murray, then of Her Majesty's 20th Regiment of Foot, writing me in November, 1851, from Montreal, where he was quartered, says :—

"Since my arrival here I have had an opportunity of speaking to Dr. French on the subject of the Count de Strzelecki's observation, and he tells me that, his attention being directed to it some time ago, he had made inquiries regarding it, and that he had obtained *positive* assertions that *squaws* frequently do have children by an Indian after bearing children to a European."

Again, Dr. Balfour, of Surinam, writing me in April, 1851, observes :—

" Living for many years in this district of South America, surrounded by a motley population of whites, negroes, and Indians, I have had very frequent opportunities of observing that Strzelecki's opinion of the effect of fruitful intercourse of the European male with the women of the aboriginal race (Indians here) is not borne out by the facts constantly occurring. The Indians of this district, known as the tribes of Warou and Arrawacka Indians, are certainly decreasing fast, but principally from the effects of that curse of the aborigines—*fire-water*. The women of these tribes are notoriously profligate during the earlier part of their life after puberty ; but after bearing children to Europeans are *frequently* fruitful with men of their own race. This is particularly observed with the women of the Warou tribe, a camp of which tribe is in the immediate

neighbourhood of the barracks, where a large detachment of European troops is constantly stationed. Any person resident in this
district could at once refer to examples of this."

And, if I mistake not, a paper was published in
the *Edinburgh Monthly Journal* in 1851 or 1852, in
which the author, either writing from the antipodes,
or giving the result of his observations while there
—whether in Australia or New Zealand, I do not
now recollect—states that *exceptional* instances, at
least, to Strzelecki's observation are of frequent
occurrence among the aboriginal females of one or
other or both of these countries.

II. Since the foregoing Essay was written, and read
to the Medical Society of Southampton, several
important communications bearing on the subject of
the inoculation theory have appeared in the *Lancet*,
from Mr. Savory, of St. Bartholomew's Hospital,[*]
Mr. Langston Parker, of Birmingham,[†] and Mr.
De Méric, of London.[‡] I purpose considering these
several communications in a future number of this
journal, and will merely remark, meanwhile, that,
by his well - conceived and admirably - executed
experiments with strychnine, Mr. Savory has
demonstrated (what Magendie and Williams failed
in demonstrating by theirs), that fluid matters do,

---

[*] "An Experimental Inquiry into the Effect upon the Mother
of Poisoning the Fœtus." *Lancet*, April 10 and 17, 1858.

[†] Lectures on Infantile Syphilis. *Lancet*, May and June,
1858.

[‡] Lettsomian Lectures—Lecture III. "On Hereditary
Syphilis." *Lancet*, September 12, 1858.

in fact, pass from the fœtus to the mother through the placenta.—Mr. Savory has thus added to the chain of evidence in support of the inoculation theory, the only link wanting to make that chain complete. Or, to put it otherwise, his discovery, for such it is, may be said to furnish the keystone of the arch on which the theory now rests, and rests securely.—(See *Essay Fifth*, pp. 130, 131.)

ESSAY FIFTH.

REMARKS ON A CASE BY THE LATE MR. HEY, OF LEEDS,
ILLUSTRATIVE OF THE TRANSMISSION THROUGH THE
FŒTUS OF SECONDARY SYPHILIS FROM THE
MALE TO THE FEMALE PARENT.

FROM
THE INDIAN ANNALS OF MEDICAL SCIENCE.
Calcutta, Vol. **xx.**, 1866.

ESSAY FIFTH.

ON THE FŒTUS IN UTERO, Etc.

In a letter to Mr. Pearson, of London, of date April 8, 1816, entitled "Facts Illustrating the Effects of the Venereal Disease on the Fœtus in Utero, and the modes of its communication,"—and published in the "Transactions of the Medico-Chirurgical Society of London" (vol. VII., part II., pp. 541 *et seq.*), Mr. Hey remarks, that it has appeared to him "from several instances, that a man may communicate the *lues venerea* after all symptoms of the disease have been removed, and he is judged to be in perfect health." And in illustration of this remark, he gives the following case :—

"A few years ago, a gentleman requested to speak with me in private. He was no sooner seated in my study, than he burst into a flood of tears, and seemed to be in the greatest distress. As soon as he could regain his composure, he informed me that he had married a lady whom he tenderly loved, and now feared that he had injured her by com-

municating the venereal disease,—which he acknowledged to have had before marriage, but from which
he believed himself to have been since that time
perfectly free. I saw no reason to doubt the truth
of his declaration.

"I visited the lady, and found her labouring
under a confirmed lues. The labia pudendi and
verge of the anus were beset with irregular fissures
and with condylomata; a discharge of puriform
matter also issued from the vagina. She was advanced to the seventh month of pregnancy; but,
before her delivery at the termination of the ninth
month, the diseased parts were healed, as I had
pursued a mercurial course with as much vigour as
seemed prudent in her condition.

" The child was at its full growth, and had no
other morbid appearance than an universal desqua-
mation of the cuticle. It continued well about a
month, and then began to grow extremely fretful,
though its evacuations indicated no disease in the
primæ viæ. At the same time, it began to have a
hoarse, squeaking voice, and soon exhibited a number
of copper-coloured blotches upon the skin. A scaly
eruption also appeared upon the chin; and the anus
showed an unnatural redness. I had no hesitation
respecting the treatment; but immediately com-
menced a mercurial course. The event was agree-
able to my wishes, and the child soon got well."

Mr. Hey then proceeds to set forth the grounds
on which he believed the affection under which both
the mother and the child laboured to be syphilis;
after which he goes on to say—"It may justly

excite surprise that the gentleman in question should have remained free from disease, when his wife was in the condition which has been described." And he " confesses himself unable to account for the circumstances."

Mr. Hey, we have seen, cites the foregoing case as one only of "*several instances*" of the kind he had met with in the course of a practice, which, at the time he wrote, had extended over a period of upwards of half a century. It is that of a man communicating the lues venerea to his wife not only after the primary disease has been cured, but after all symptoms of the disease have been removed, and he is judged to be in perfect health.

It is to this point that I wish to call attention. There is here no question as to whether in its *overt* or *manifest* form, secondary syphilis is transmissible from the male to the female sex,—whether, for example, a man actually ill of that form of syphilis, and seen of all men to be so, may infect his wife. The case just now in view is that of a man once affected with syphilis, imparting the disease to his wife *after all traces of it have disappeared from him*, contracting marriage at a time when, to all appearance, he is so entirely free of the poison as to be entitled to a " clean bill of health," and yet in due season, without again exposing himself to re-infection, infecting his wife.

Mr. Hey offers no explanation of this class of cases. He seems to take it for granted—manifestly he does so—that the man communicates the disease to his wife *directly*. For, as to the case in hand, he expresses surprise that, after she had herself become diseased to the extent described by him, the lady did not, in her turn, re-infect her husband,—now (in his view) capable of re-infection. It puzzles him to think how the man with no apparent trace about him of the virus, yet able notwithstanding to impart it to his wife, should in his seemingly sound state have remained unaffected by her, she having the labia pudendi and anus beset with fissures and condylomata, and the vagina the seat of a puriform discharge. It is the husband's immunity from re-infection, not the wife's susceptibility of the virus, that gives the case its main interest in Mr. Hey's eyes: the man apparently free of the virus, yet infecting his wife; . the wife "labouring under a confirmed lues," yet failing to re-infect her husband ! It is the marked contrast between the two in this respect, that excites his surprise. But he raises no question as to how it came about that the gentleman contaminated his wife.

At the time Mr. Hey wrote (1816), no other explanation of the transmission of the virus in such cases, seems to have been thought of than this— namely, that it is somehow accomplished through the agency of the man's secretions on the woman's blood into which they are received,—or on the mucous surfaces of her body to which they are applied,—whether of the seminal fluid in coitu, or of the mucous secretions, either in coitu or in kissing.

It is only of late years, in fact, that another mode of transmission has been suggested,—that, namely, through the fœtus—itself begotten of the father, from him inheriting the virus, and having *in utero* the disease resulting from it in such manner and degree as to impart it to its mother, through the medium of the placenta.

If this view be correct, then it is not that the husband directly contaminates his wife in any of the ways formerly supposed, but that he begets a syphilitic child,—which child it is that really contaminates its mother. It is not his having sexual intercourse with his wife that infects her, but his begetting a syphilitic child within her body. Impregnation, therefore, and not mere coitus, is the essential condition of the conveyance of the poison from the man to the woman. It is the fœtus that is the real transmitter of the poison. In Mr. Hey's case this condition obtained. When consulted regarding it, Mr. Hey found the lady "advanced to the seventh month of pregnancy." And if in this case it was in the way just indicated that the lady imbibed the virus and became diseased; and if, further, secondary syphilis is rarely (not to say never), transmitted from husband to wife in any other way, it will excite less surprise—or no surprise at all, that the gentleman escaped re-infection, although cohabiting with his diseased wife.

Some of my readers may possibly recollect that, several years ago, Professor Harvey, of Aberdeen, directed attention to the power of the fœtus to inoculate the maternal with the peculiarities of the paternal organism; and that he then suggested the probability of secondary syphilis being imparted in this way.*

Since then Professor Paget, Mr. Jonathan Hutchinson, and others in England ; Dr. J. B. Balfour and Dr. Matthews Duncan, in Scotland ; and MM. Ricord and Diday, in France,—have been working in this field of inquiry; and they have established, as a matter of fact, what Dr Harvey could only then indicate as probable. Long previously, however, Dr. Ramsbotham, first of all (1835), and subsequently to him, Dr. Montgomery (1837), had given expression to the very same idea. Nothing can be clearer than Dr. Ramsbotham's own statement both of the idea itself and of the grounds on which he reached it : indeed, no more precise exposition of either has yet been given by any that have come after him.

"It is a generally received opinion, I believe, (says Dr. Ramsbotham) that syphilis in its secondary stage is not communicable directly to either sex from the other,—that the disease is not propagated unless there exist an open chancre ; and *this accords with my observation*. But it appears to me *probable* that, if a previously healthy woman conceive of an ovum tainted by syphilitic virus derived from its father, her system may become *inoculated during the progress of gestation*, in consequence of the close vascular connection existing between it and herself. For, it has fallen to my lot to see *more*

---

* Edinburgh *Monthly Journal of Medical Science*, October, 1849, and October and November, 1850.

than one case, in which a young woman united to a man labouring under *obstinate* secondary symptoms " (a much stronger case this than Mr. Hey's) " remained *healthy* for *some months after marriage,* but became the subject of the same disease in its secondary form *soon after impregnation had taken place ;* and I have considered that, in such a case, the mother derived the disease *not directly* from the *father,* but from the *affected infant* which she carried in her womb."*

With regard to the general principle, so happily expressed by Dr. Ramsbotham, namely this, " that, in the class of cases in question, the woman derives the syphilitic virus not directly from the man, but from the infected infant begotten by him in her womb," Professor Harvey has adduced such proof as seems to me decisively to establish it.†

The proof is of this sort : proof, on the one hand, that so long as the woman remains *unimpregnated,* she may cohabit with a man affected with secondary syphilis even in an overt or palpable form, and escape contamination ; proof, on the other hand, that, in the event of *impregnation occurring,* she may and often does become contaminated, even when the disease exists in her husband in a latent and unsuspected form :—the contamination showing itself commonly

* *Medical Gazette,* May 23rd, 1835.

† " On the fœtus in utero as inoculating the maternal with the peculiarities of the paternal organism ; and on the transmission thereby of secondary or constitutional syphilis from the male to the female parent." *Glasyow Medical Journal,* January, 1859.

about the third or fourth month of pregnancy,—this manifestation of it subsiding shortly after delivery, and re-appearing on the woman again becoming pregnant. This kind of proof, copious as it now is, is, I think, peculiarly satisfactory. It indicates *freedom* from contamination under circumstances which seem peculiarly favourable to its occurrence,— the husband being *sensibly* affected by the poison, and *actual* contamination under circumstances seemingly peculiarly unfavourable,—the husband presenting no traces of the virus on any part of his person. The inference appears unavoidable—that (in a large proportion of cases at least) impregnation is the essential condition of the transmission of the virus from the man to the woman ; that the child begotten by the man and tainted by him, is the proximate channel whereby the poison is conveyed to the woman,—the real transmitter of it from the father to the mother. Small—infinitesimally small —the amount may be of the virus in the father's blood,—yet enough to grow and to multiply itself (after the manner of a ferment) in the tender and plastic organism of the fœtus inheriting it : enough to taint it, and, as actively developed in it, to taint the mother.

Mr. Savory, of St. Bartholomew's, may well be said, I think, to have supplied " the keystone to the arch " on which the proof of this view of maternal contamination now rests. By a series of well-devised and happily - executed experiments, he has succeeded in demonstrating what Magendie, Williams, and other experimenters, had failed to show,—that fluid matters may and do, in fact, pass from the fœtus to the mother through the *placenta.* His experiments consisted in introducing a solution of strychnia, of known strength, into the circulating system of the fœtus while still connected with the mother by the cord. The result was, the death of the mother, as well as of the fœtus—the death of the mother taking place in a mode that clearly implied the action of strychnia on her system, and under circumstances that unequivocally indicated the transmission of the poison along the placenta from the fœtus into her blood.*

Having, in this paper, a special object in view, which is, to offer some remarks on Mr. Hey's case, I content myself with this *general* statement as to the *kind* of evidence we now have for the belief, that the fœtus is *one* channel, and in most cases, the *special* channel, whereby secondary syphilis passes from the male to the female parent : and I purposely forbear adducing the *details* of the evidence supplied by a not inconsiderable number of independent observers. But I should wish it to be understood that I do not allege or maintain that the fœtus is the only channel of transmission. Dr. Ramsbotham, we have seen, thinks that secondary syphilis is not communicable *directly,* and this opinion he holds in common with many. It would appear, however, that it is thus directly communicable. Facts at least there are which seem to indicate this. Still, the

---

* "An experimental inquiry into the effect upon the mother of poisoning the fœtus."—*Lancet,* April 10 and 17, 1858.

weight of evidence is, I think, strongly in favour of
the presumption that the direct transmission of the
virus is, *comparatively*, of rare occurrence. Whether,
as thus transmissible, a solution of continuity of some
kind—(ulceration, abrasion, fissure,) is essential
thereto, does not clearly appear; probably it is.
But I am well satisfied of this, namely, that in as
far as the disease is directly communicable, it is
only as existing in a well-marked and palpable form,
i.e., when the husband is *unmistakably* labouring
under it. When existing in a very mild, or (as in
Mr. Hey's case) in a latent form, I do not believe
it to be transmissible otherwise than through the
foetus.

How it is that it should be so readily conveyed to
the mother through *this* particular channel, one is
not called upon to say. It were enough to establish
it as a matter of fact. But a not unreasonable explanation of it may be given. An amount of the
virus in any of the secretions of the man that might
be inadequate to hurt the full-grown woman, may,
as existing in his semen, be sufficiently potent to
taint the embryonic cell fertilized in her womb.
That constituent of the semen which, infinitesimal
as it is even in its totality, is adequate to the production of a *new being*, and to the implanting in it
all the qualities, physical and moral, of the father,
may contain enough of the virus to bear fruit,
after its kind, in the foetus. In the delicate tissues
in the course of evolution within the embryo, and
in the " *tourbillon continuel* " of the vital changes
which for nine months the embryo is, in a very

special sense, the seat of, the virus may find conditions peculiarly favourable to its own growth and increase; and taking full effect first on the susceptible organism of the fœtus, it may there acquire force enough and volume enough to affect and contaminate the mother also. "Behold how great a matter a little fire kindleth." Whole hosts of seminal animalcules, as numerous "as there were men in the army of Xerxes," operating for weeks and months and years together, might prove powerless against the woman; and yet the one tainted animalcule that enters the breach in the maturated ovum in which it disappears, may be mighty enough, through the train of events which follow on its lodgement there, to the spoiling of the mother. This by the way, however, *valeat quantum.* The fact remains, that secondary syphilis, rarely transmissible to the female *directly,*—and then only (I believe) when existing in the husband in an overt form, is readily transmitted by the fœtus, even when existing in the father in a mild, or in a latent and unsuspected form.

Perhaps, I may best convey to my readers what I believe to be the true state of the case, if I say, that, in the event of syphilis existing in a *palpable* form in the husband, there are ninety-nine chances to one in favour of the woman's remaining healthy,—so long as she remains unimpregnated,—and ninety-nine chances to one against her, on her falling with child; while, if the husband have the disease in a *latent* form, she runs no risk at all, so long as she *escapes* impregnation.

I pass on now to offer some remarks on Mr. Hey's case.

It was not with any view to endorse the theory of the fœtal transmission of secondary syphilis that the case in question was given by Mr. Hey. This theory he had probably no idea of : it was not in his day a doctrine of the schools. His allusion to the lady being "advanced to the seventh month of pregnancy" comes in *incidentally.* Had he omitted to mention it, or to speak of the child, the case would have been valueless in relation to that theory. Yet the reference to the pregnancy and to the child, had no particular bearing on the object Mr. Hey had in view. The pregnancy comes in as a fitting prelude to what he had to say of the child, and the child as supplying corroborative evidence, by its own syphilitic condition, of the syphilitic condition of the mother,—the whole value of the case, in Mr. Hey's view, lying in the marvellous escape of the husband from re-infection. As the case stands, however, it has a value of another sort, not contemplated by Mr. Hey.

Nor is it as furnishing a decisive *proof* of this doctrine of inoculation, that I am myself disposed to attach value to the case. It will presently appear that it is so far defective for such a purpose. It is simply as a case (casually picked up, among others like it, in the course of my reading,) which *fits* into that doctrine, that I adduce it. I have no doubt, however, in my own mind, that it was a genuine instance of fœtal inoculation, though I ground nothing on it in the way of proof. Looking at it in the clear light reflected on it by the facts by which the doctrine is now decisively established, I cannot but believe that the lady *was* contaminated by her own child *in utero*. The case is of a piece with that of many others, now on record, in which the evidence of such contamination was clear and unequivocal. The gentleman had once been affected with syphilis, and was quite aware of the circumstance. But it was before his marriage that he had it, and ere this all traces of it had passed away. His connection with his wife was fruitful; she become pregnant, and, in the seventh month, was found by Mr. Hey to be "labouring under a confirmed lues,"—a condition, the true nature of which was in due time verified by the birth of a syphilitic infant.

One link, indeed, in the chain of evidence against the fœtus, is wanting. To make the proof against it complete, it would require to be shown that the mother kept free of all syphilitic affection until she *fell with child*. Mr. Hey being dead, this missing link cannot now be supplied. Perhaps, even at the time he had the case in hand, it would not have been possible for him to supply it. For, it is conceivable (indeed the narrative almost carries the supposition on the face of it) that the marriage was a recent one; and it is quite possible that the impregnation was *contemporaneous* with its celebration. If so, as in any such case, there would be a plain source of fallacy in the matter. But making due allowance for this (and making no use, because of it, of the case in the way of proof) there are, I think, fair grounds for connecting the lady's contamination with impregnation. The manifestation of the disease in her could not well have been of long standing, before Mr. Hey was consulted. The behaviour of the husband, as recorded by Mr. Hey, makes it probable that he would not have allowed his wife to remain four, or five, or six months ill in the way he suspected, without seeking advice; and she was already seven months gone with child. The probability is, that Mr. Hey was called in before the disease had existed many weeks,—in which case we should have warrant enough for our inference.

In short, admitting again that there is no *direct* or *absolute* proof of the contamination being the result of impregnation, I cite the case as one that affords a good illustration of the theory of inoculation by and through the fœtus. Believing, myself, that the lady derived the infection from the child *in utero*, I shall not quarrel with anyone who sees the case in a different light. It was not given by Mr. Hey with any view to the inoculation theory. I look upon it as what the late Dr. Blunt of Cambridge would have termed an "undesigned coincidence." It fits, *unintentionally*, into the inoculation theory; and coming upon it (as I remarked before) casually and accidentally, and looking at it in the face of facts discovered since it was published, I cannot help seeing in it a singularly striking coincidence of the undesigned kind, in relation to that theory.

One word as to another feature of this case—that, seemingly, of chief interest to Mr. Hey. That the lady, with "her labia and anus beset with fissures and condylomata, and the vagina the seat of a puriform discharge," should have failed to re-infect her husband, is to Mr. Hey a matter of surprise. " It may justly excite surprise (he remarks) that the gentleman in question should have remained free from disease when his wife was in the condition which has been described." And he confesses himself unable, without calling in the aid of a supposition which he says " wants probability," to account for the circumstance.

This supposition Mr. Hey does not even indicate. Doubtless, what he had in view was, that subsequently to the lady becoming diseased, or sufficiently diseased to infect her husband *per coitum*, the parties had abstained from all sexual intercourse. This is not likely: it is a supposition which " wants probability." Very likely, after the gentleman's suspicions were fully roused, there may have been no intercourse between them; but prior to this—if infection could have come to him in that way, there was probably disease enough in her genitals to infect him, and intercourse enough to insure infection. This we may guess was the " supposition " Mr. Hey had in his eye. Mr. Hey marvels how it could be that he, tainted indeed, yet so slightly as to be apparently free of taint—with no disease on his genitals, with none anywhere except in his blood, and, peradventure, in his secretions, should infect his wife *per coitum;* while she, on her part, notably diseased, should in that way be incapable of infecting or re-infecting him.

Two remarks occur to me in connection with this

point. The *first* is, that had Mr. Hey been aware of
the theory of inoculation, and a believer in it:—had
he been alive to the possibility, and the probability,
of the lady having derived the infection through
the medium of the fœtus ; and had he known the
rarity with which secondary syphilis, even in its
well-marked form, is ever transmitted from the
male to the female, except when impregnation
follows intercourse, he would probably have seen
no cause for surprise in the matter. As it was not
per coitum that the gentleman contaminated his
wife, so there could be nothing marvellous in the
latter not re-infecting him.

My next remark is, that, apart from any consideration of this kind, there is room for doubt (to
say the least) whether in any case *re-infection* is
possible,—possible at all events so long as one con-
tinues under the influence of the virus—contracted
in the ordinary way. The gentleman in question
having once had syphilis, and being still tainted by
it, it is a fair presumption that in the circumstances
he was incapable of re-infection.

In bringing this paper to a close, I cannot forbear
adducing a case (hitherto unrecorded) bearing on
the theory of fœtal inoculation,—a case communicated to my father by his colleague Dr. Dyce,
Professor of Midwifery in the University of Aberdeen, and by him to me. It is that of a medical
man acquiring syphilis, in its primary form, on the
forefinger of the right hand, from a woman attended
by him in labour, and affected with the disease in
that form—(the finger being abraded at the time) ;

and subsequently contaminating his wife. A well-marked chancre formed on the finger, and in due time constitutional symptoms followed. Some time after, this gentleman's wife became the subject of constitutional syphilis—yet *not until she fell with child.* A period of many months intervened between the infection of the husband and the subsequent infection of the wife, during all which time sexual intercourse obtained, but no impregnation. Impregnation eventually occurred, and the manifestation of syphilis in the lady (who had the disease in its secondary form only) were coincident with the manifestations of pregnancy.

I may just add, as to this case, on the authority of my father, that the parties concerned were intimately known to Dr. Dyce, and that he was personally conversant with the whole facts; and further that Dr. Dyce entertained no doubt whatever, that the lady derived the infection through the medium of the fœtus, and in no other way.

I have already stated, once and again, that I merely adduce Mr. Hey's case as one that *tallies* with the theory of inoculation through the fœtus, and that serves to *illustrate* that theory. What it wants in the way of *proof* is the omission of all reference to the precise time when the first manifestations of syphilis presented themselves in the lady; she was advanced to the seventh month of pregnancy when seen by Mr. Hey; but she may have conceived on the occasion of the first *coitus* following her marriage; her infection may have been concurrent with the coitus; and possibly (although it does not seem likely) the disease may have broken out in her long before Mr. Hey was consulted.   Dr. Dyce's case, however, is wanting in nothing that it concerns us to know :—In it, the infection of the lady was coincident with her impregnation; mere coitus carried on for months together, was harmless to her.   And the case may confidently be cited as one of the kind, now sufficiently numerous, that *decisively establish* the theory in question.

LONDON : H. K. LEWIS, 136, GOWER

BY THE SAME AUTHOR.

I.

FIRST LINES OF THERAPEUTICS:
AS BASED ON THE MODES AND PROCESSES OF HEALING AS OCCURRING SPONTANEOUSLY IN DISEASE,
AND
ON THE MODES AND PROCESSES OF DYING AS RESULTING NATURALLY FROM DISEASE.
1879.   Price 5s.

LONDON: H. K. LEWIS.

II.

TREES AND THEIR NATURE;
OR, THE BUD AND ITS ATTRIBUTES.
In a Series of Letters to his Sons.   Illustrated with Engravings.
1856.   Price 5s.

LONDON: JAMES NISBET & CO.

III.

THE TESTIMONY OF NATURE TO THE IDENTITY OF THE BUD AND THE SEED.
(Supplementary to the foregoing.)
1857.   Price 2s. 6d.

LONDON: JAMES NISBET & CO.

IV.

GOOD THE FINAL GOAL OF ILL;
OR, THE BETTER LIFE BEYOND.
IN FOUR LETTERS TO THE VEN. ARCHDEACON FARRAR.
1883.   Price 3s. 6d.

LONDON: MACMILLAN & CO.

V.

MAN'S PLACE AND BREAD UNIQUE IN NATURE, AND HIS PEDI-GREE HUMAN, NOT SIMIAN.
1865.   Price 1s.

EDINBURGH: EDMONSTON & DOUGLAS.

SELECTED LIST OF WORKS

PUBLISHED BY

H. K. LEWIS,

136, GOWER STREET, LONDON, W.C.

₊ *Complete Catalogue post free on application.*

Medium 8vo, 21s.

CONTRIBUTIONS TO PATHOLOGY
AND THE
PRACTICE OF MEDICINE.

By JOHN R. WARDELL, M.D., F.R.C.P.,
Late Consulting Physician to Tunbridge Wells General Hospital.

Third Edition, Revised and Enlarged, with numerous Illustrations, 8vo, 18s.

THE

SCIENCE & ART OF MIDWIFERY.

By WILLIAM THOMPSON LUSK, A.M., M.D.,
Professor of Obstetrics and Diseases of Women in the Bellevue Hospital Medical College, &c.

Eleventh Edition, demy 8vo, 15s.

A HANDBOOK OF THERAPEUTICS.

By SYDNEY RINGER, M.D., F.R.S.,
Professor of Medicine in University College, Physician to University College Hospital, &c.

With 30 Plates from Original Drawings, of which 28 are Coloured, and numerous Wood Engravings, demy 8vo, 14s.

AN INTRODUCTION TO PRACTICAL BACTERIOLOGY.

BASED UPON THE METHODS OF KOCH.

By EDGAR M. CROOKSHANK, M.B., Lond.,
Demonstrator of Physiology, King's College, London.

Works Recently Published by H. K. Lewis.

Fourth Edition, Revised, limp roan, med. 24mo, 7s.

THE EXTRA PHARMACOPŒIA.

WITH THE ADDITIONS INTRODUCED INTO THE BRITISH PHARMA-
COPŒIA, 1885; AND MEDICAL REFERENCES, AND A THERA-
PEUTIC INDEX OF DISEASES AND SYMPTOMS.

By WILLIAM MARTINDALE, F.C.S.,

*Late Examiner of the Pharmaceutical Society, and late Teacher of Pharmacy
and Demonstrator of Materia Medica at University College,*

AND

W. WYNN WESTCOTT, M B., Lond.,

Deputy Coroner for Central Middlesex.

Seventh Edition, Edited and Revised by ROBERT J. HESS, M.D.,
with twelve full-page plates, five being coloured, and 165 Wood
Engravings, 1,081 pages, royal 8vo, 35s.

OBSTETRICS: *The Theory and Practice;*

INCLUDING THE DISEASES OF PREGNANCY AND PARTURITION,
OBSTETRICAL OPERATIONS, &C.

By P. CAZEAUX,

Adjunct Professor in the Faculty of Medicine of Paris, &c.,

AND

S. TARNIER,

*Professor of Obstetrics and Diseases of Women and Children in the Faculty of
Medicine of Paris.*

Royal 8vo, 18s.

THE PHYSIOLOGY OF THE SPECIAL SENSES AND GENERATION;

(Being Vol. V. of the Physiology of Man).

By AUSTIN FLINT, Jun., M.D.,

*Professor of Physiology and Physiological Anatomy in the Bellevue Medical Col-
lege, New York; attending Physician to the Bellevue Hospital, &c.*

Works Recently Published by H. K. Lewis.

Sixth Edition, with Illustrations, in one volume of over 1,000 pages, large 8vo, 21s.

A HANDBOOK OF THE THEORY & PRACTICE OF MEDICINE.

By FREDERICK T. ROBERTS, M.D., B.Sc., F.R.C.P.

Examiner in Medicine at the Royal College of Surgeons; Professor of Therapeutics in University College; Physician to University College Hospital; Physician to the Brompton Consumption Hospital, &c.

Fcap. 8vo, 7s. 6d.

NOTES ON MATERIA MEDICA AND PHARMACY.

By THE SAME AUTHOR.

Small 8vo, pp. 288, 6s.

ROME IN WINTER & THE TUSCAN HILLS IN SUMMER.

A CONTRIBUTION TO THE CLIMATE OF ITALY.

By DAVID YOUNG, M.C., M.B., M.D.,

Licentiate of the Royal College of Physicians, Edinburgh; Licentiate of the Royal College of Surgeons, Edinburgh; late Professor of Botany in the Grant Medical College, Bombay; Fellow of, and late Examiner in Midwifery to, the University of Bombay; and Fellow of the Obstetrical Society of London.

With Illustrations, crown 8vo, 10s. 6d.

A PRACTICAL TREATISE ON

DISEASES OF THE KIDNEYS AND URINARY DERANGEMENTS.

By C. H. RALFE, M.A., M.D. Cantab., F.R.C.P. Lond.,

Assistant Physician to the London Hospital; late Senior Physician to the Seamen's Hospital, Greenwich.

LONDON : H. K. LEWIS, 136, GOWER STREET, W.C.

www.ingramcontent.com/pod-product-compliance
Lightning Source LLC
Chambersburg PA
CBHW020558270326
41927CB00006B/896